Toxicological and Performance Aspects of Oxygenated Motor Vehicle Fuels

Committee on Toxicological and
Performance Aspects of
Oxygenated Motor Vehicle Fuels

Board on Environmental Studies and Toxicology
Commission on Life Sciences
National Research Council

NATIONAL ACADEMY PRESS
Washington, D.C. 1996

NATIONAL ACADEMY PRESS
2101 Constitution Ave., N.W., Washington, D.C. 20418

NOTICE: The project that is the subject of this report was approved by the Governing Board of the National Research Council, whose members are drawn from the councils of the National Academy of Sciences, the National Academy of Engineering, and the Institute of Medicine. The members of the committee responsible for the report were chosen for their special competences and with regard for appropriate balance.

This report has been reviewed by a group other than the authors according to procedures approved by a report review committee consisting of members of the National Academy of Sciences, the National Academy of Engineering, and the Institute of Medicine.

The project was supported by the U.S. Environmental Protection Agency under contract no. 68D50054.

Library of Congress Catalog Card No. 96-69352
International Standard Book No. 0-309-05545-8

Additional copies of this report are available from:

National Academy Press
2101 Constitution Ave., NW
Box 285
Washington, DC 20055
800-624-6242
202-334-3313 (in the Washington Metropolitan Area)

Printed in the United States of America

Committee on Toxicological and Performance Aspects of Oxygenated Motor Vehicle Fuels

BAILUS WALKER *(Chair)*, Howard University Cancer Center, Washington, D.C.
ROBERT C. BORDEN, North Carolina State University, Raleigh, N.C.
WILLIAM S. CAIN, University of California, San Diego, Calif.
STEVEN D. COLOME, Integrated Environmental Services, Irvine, Calif.
DAVID B. COULTAS, University of New Mexico Health Sciences Center, Albuquerque, New Mex.
W. ROBERT EPPERLY, Catalytica Advanced Technologies, Inc., Mountain View, Calif.
CHARLES H. HOBBS, Inhalation Toxicology Research Institute, Albuquerque, New Mex.
SIMONE HOCHGREB, Massachusetts Institute of Technology, Cambridge, Mass.
JOHN H. JOHNSON, Michigan Technological University, Houghton, Mich.
DOUGLAS R. LAWSON, Colorado State University, Fort Collins, Colo.
ERNEST E. MCCONNELL, Raleigh, N.C.
SANDRA N. MOHR, Environmental and Occupational Health Sciences Institute, Piscataway, N.J.
PHILLIP S. MYERS, University of Wisconsin, Madison, Wis.
JOSEPH V. RODRICKS, ENVIRON International Corp., Arlington, Va.
DONNA SPIEGELMAN, Harvard School of Public Health, Boston, Mass.

Staff

CAROL A. MACZKA, Project Director and Program Director, Toxicology and Risk Assessment Program
RAYMOND A. WASSEL, Program Director, Environmental Sciences and Engineering Program
JAMES J. ZUCCHETTO, Director, Board on Energy and Environmental Systems
CATHERINE M. KUBIK, Senior Project Assistant
RUTH P. DANOFF, Senior Project Assistant
KATHRINE IVERSON, Library Assistant

Sponsor

U.S. Environmental Protection Agency

Board on Environmental Studies and Toxicology

Staff

JAMES J. REISA, Director
DAVID J. POLICANSKY, Associate Director and Program Director for Natural Resources and Applied Ecology
CAROL A. MACZKA, Program Director for Toxicology and Risk Assessment
LEE R. PAULSON, Program Director for Information Systems and Statistics
RAYMOND A. WASSEL, Program Director for Environmental Sciences and Engineering

Commission on Life Sciences

Other Recent Reports of the Board on Environmental Studies and Toxicology

Carcinogens and Anticarcinogens in the Human Diet: A Comparison of Naturally Occurring Synthetic and Natural Substances (1996)
Upstream: Salmon and Society in the Pacific Northwest (1996)
Science and the Endangered Species Act (1995)
Wetlands: Characteristics and Boundaries (1995)
Biologic Markers (Urinary Toxicology (1995), Immunotoxicology (1992), Environmental Neurotoxicology (1992), Pulmonary Toxicology (1989), Reproductive Toxicology (1989))
Review of EPA's Environmental Monitoring and Assessment Program (three reports, 1994-1995)
Science and Judgment in Risk Assessment (1994)
Ranking Hazardous Sites for Remedial Action (1994)
Pesticides in the Diets of Infants and Children (1993)
Issues in Risk Assessment (1993)
Setting Priorities for Land Conservation (1993)
Protecting Visibility in National Parks and Wilderness Areas (1993)
Dolphins and the Tuna Industry (1992)
Hazardous Materials on the Public Lands (1992)
Science and the National Parks (1992)
Animals as Sentinels of Environmental Health Hazards (1991)
Assessment of the U.S. Outer Continental Shelf Environmental Studies Program, Volumes I-IV (1991-1993)
Human Exposure Assessment for Airborne Pollutants (1991)
Monitoring Human Tissues for Toxic Substances (1991)
Rethinking the Ozone Problem in Urban and Regional Air Pollution (1991)
Decline of the Sea Turtles (1990)

These reports may be ordered from the National Academy Press (800) 624-6242; (202) 334-3313

Preface

In response to public concerns over the use of methyl tertiary-butyl ether (MTBE)-oxygenated fuels in motor vehicles, a comprehensive assessment of such fuels was drafted by the federal government under the direction of the White House Office of Science and Technology Policy (OSTP). The assessment addressed public health, air quality, water quality, fuel economy, and engine performance.

This National Research Council report independently reviews the federal assessment in terms of scientific credibility, comprehensiveness, and internal consistency of the information presented. We hope that OSTP, in preparing its final report, will provide specific responses to the conclusions and recommendations provided in this report.

As part of its information gathering, our committee heard presentations from the principal authors of the various documents that composed the interagency report. Carleton J. Howard, of the U.S. Department of Commerce, summarized the document on Air Quality Benefits; Arthur C. Upton, chairman of the Health Effects Institute's Oxygenates Evaluation Committee, and Mary White, of the U.S. Department of Health and Human Services, discussed the documents assessing the potential health risks associated with oxygenated gasoline; John S. Zogorski, of the U.S. Department of the Interior, discussed the document on fuel oxygenates and water quality; and David J. Kortum, of the U.S. Environmental Protection Agency (EPA) presented the document on fuel economy and engine performance issues. Robert T. Watson, associate director for environment, OSTP, provided the committee with an overview of the interagency effort. Two representatives of EPA, the sponsor of this National Research Council study, provided useful information and perspectives for the committee: William H. Farland, director, national center for environmental assessment, and Mary D. Nichols, assistant administrator for air and radiation. The committee also heard presentations by Larry S. Andrews, ARCO Chemical Company; Robert G. Tardiff, EA Engineering, Science, and Technology, Inc.; and Bernard D. Goldstein and Paul J. Lioy, Environmental and Occupational Health Sciences Institute.

We are grateful for the assistance of the National Research Council staff in the preparation of this report. In particular, the committee wishes to acknowledge Carol A. Maczka, project director, whose hard work and expertise were most effective in bringing this report to completion. Also, Raymond A. Wassel, program director for environmental sciences and engineering, and James J. Zucchetto, director of the Board on Energy and Environmental Systems, provided the committee with outstanding technical assistance in the preparation of this report. Other staff members who contributed to this effort are Paul Gilman, executive director of the Commission on Life Sciences; James J. Reisa, director of the Board on Environmental Studies and Toxicology; and Catherine M. Kubik, senior project assistant.

Last, but by no means least, the work of all the members of the committee is greatly appreciated.

Bailus Walker, Jr., *Chair*
Committee on Toxicological and Performance Aspects of Oxygenated Motor Vehicle Fuels

The National Academy of Sciences is a private, nonprofit, self-perpetuating society of distinguished scholars engaged in scientific and engineering research, dedicated to the furtherance of science and technology and to their use for the general welfare. Upon the authority of the charter granted to it by the Congress in 1863, the Academy has a mandate that requires it to advise the federal government on scientific and technical matters. Dr. Bruce Alberts is president of the National Academy of Sciences.

The National Academy of Engineering was established in 1964, under the charter of the National Academy of Sciences, as a parallel organization of outstanding engineers. It is autonomous in its administration and in the selection of its members, sharing with the National Academy of Sciences the responsibility for advising the federal government. The National Academy of Engineering also sponsors engineering programs aimed at meeting national needs, encourages education and research, and recognizes the superior achievements of engineers. Dr. Harold Liebowitz is president of the National Academy of Engineering.

The Institute of Medicine was established in 1970 by the National Academy of Sciences to secure the services of eminent members of appropriate professions in the examination of policy matters pertaining to the health of the public. The Institute acts under the responsibility given to the National Academy of Sciences by its congressional charter to be an adviser to the federal government and, upon its own initiative, to identify issues of medical care, research, and education. Dr. Kenneth I. Shine is president of the Institute of Medicine.

The National Research Council was organized by the National Academy of Sciences in 1916 to associate the broad community of science and technology with the Academy's purposes of furthering knowledge and advising the federal government. Functioning in accordance with general policies determined by the Academy, the Council has become the principal operating agency of both the National Academy of Sciences and the National Academy of Engineering in providing services to the government, the public, and the scientific and engineering communities. The Council is administered jointly by both Academies and the Institute of Medicine. Dr. Bruce Alberts and Dr. Harold Liebowitz are chairman and vice chairman, respectively, of the National Research Council.

CONTENTS

Executive Summary

Carbon monoxide (CO) pollution is mainly caused by incomplete combustion of motor-vehicle fuels. The Clean Air Act Amendments of 1990 require the use of oxygenated gasoline in areas of the country where the National Ambient Air Quality Standard (NAAQS) for CO is being exceeded. This requirement was intended to protect those who are most susceptible to the adverse health effects of CO, particularly patients with cardiovascular disease. Methyl tertiary-butyl ether (MTBE) has become the most widely used oxygenate in the United States for motor-vehicle fuels. (Other oxygenates used in fuels include ethanol, tertiary-butyl alcohol, ethyl tertiary-butyl ether, and tertiary-amyl methyl ether.)

Concurrently with the start of the federal oxygenated-gasoline program in 1992, MTBE has been implicated in complaints of headaches, coughs, and nausea, principally in Alaska, but also in Montana, New Jersey, and Wisconsin. There have also been anecdotal reports of reduced fuel economy in some locations and questions about engine performance. Additional concerns have been raised about the detection of low levels of MTBE in some samples of groundwater. Due to the public concern over the potential health effects of MTBE-oxygenated fuels, more than $2 million of scientific studies have been conducted by EPA and others to investigate the reported symptoms. Unfortunately, all of these studies have deficiencies, such as inadequate exposure assessment, insufficient sample size, subjective outcome assessment, and the possibility of selection bias.

EPA asked the National Research Council to independently review a draft interagency report from the federal government, prepared under the direction of the Office of Science and Technology Policy (OSTP) through the Committee on Environment and Natural Resources (CENR) of the president's National Science and Technology Council (NSTC). The purpose of the interagency report was to provide a review of the scientific information on oxygenated fuels and to assess effects of the winter oxygenated-fuels program on air quality, fuel economy, engine performance, water quality, and public health.

In response to EPA's request, the Research Council assembled a multidisciplinary committee—the Committee on Toxicological and Performance Aspects of Oxygenated Motor Vehicle Fuels—which has prepared this report. The committee was charged with the following tasks: (1) review the draft interagency report on the toxicological effects of MTBE and other oxygenates and compare such effects and multimedia exposures with those of conventional gasoline; (2) review the draft interagency report on the impacts of MTBE and other oxygenates on vehicle emissions, air pollution, fuel economy, and engine performance and compare such impacts with those of conventional gasoline; and (3) identify priorities for research to fill data gaps. The committee was not provided with the original sources of data from which the interagency report was written, so a full and complete critique of scientific credibility, comprehensiveness, and internal consistency of the data was not possible within the constraints of this study. Although the charge to the committee did ask for a comparison of oxygenated fuels with conventional gasoline, it was in the context of reviewing the interagency report. The comparison was not possible, because the interagency document failed to make such comparisons with respect to health effects, in large part due to the limited amount of comparative data. Comparative results were available in the interagency report on air quality, fuel economy, and engine performance and were considered by the committee. The committee was asked to address only the interagency report, which dealt with the winter oxygenated-fuels program, not the program itself. In this report, "oxygenated fuels" is intended to refer only to the fuels used within the oxygenated-fuels program. The interagency report did not specifically examine the reformulated gasoline program, which is intended to reduce motor-vehicle emissions that lead to increased ozone concentrations in the lower atmosphere.

The committee's critique of the interagency report submitted to the NRC on March 15, 1996, and recommendations for further studies are summarized below. (The preface and executive summary of the interagency report are presented in Appendix A.)

AIR QUALITY

CRITIQUE OF INTERAGENCY DRAFT REPORT

The interagency report concludes that there have been substantial reductions in ambient CO concentrations in the past 20 years and that vehicle emission controls have been a major factor in this reduction. The committee agrees with these conclusions. However, the committee believes that the federal report should better characterize the uncertainty about the extent to which oxygenated fuels have contributed to this reduction. Although the interagency report provides a good summary of previous studies that have assessed the impacts of oxygenated fuels on winter CO concentrations, it should state clearly that winter ambient CO reductions attributable to oxygenated fuels have been as low as zero and as high as about 10%. In addition, the implications of emission-data deficiencies in evaluating the effectiveness of the oxygenated-fuels program should be fully discussed in the interagency report.

Because of the relationship between ambient temperature and tailpipe emissions of CO, it is important that the term "winter temperature" be well defined in the interagency report. A map, or similar mechanism, giving winter temperature ranges and mean temperatures for the areas participating in the oxygenated-fuels program is essential in understanding and evaluating the data.

The committee agrees with the interagency report that, under many fuel-control systems, oxygenated fuels decrease CO emissions under Federal Test Procedure conditions (at 75°F). However, the data presented do not establish the existence of this benefit under winter driving conditions. Also, the report does not clearly address the effects of fleet composition, particularly high-emitting vehicles, on total CO emissions. Emissions from high emitters can be two orders of magnitude larger than those from late model, well-maintained vehicles.

EPA designed a computer model, referred to as the MOBILE model, specifically for use by states in preparing emission inventories required under the Clean Air Act. MOBILE 5a is the model version currently in use. The interagency report highlights some discrepancies between MOBILE 5a model results, vehicle-emissions data, and ambient concentrations. However, it does not provide an assessment of why those differences exist. It should do so and should also emphasize to a greater extent that the model apparently overpredicts the oxygenated-fuel effect by at least a factor of 2 based on comparisons of model predictions of CO emission reductions with observed data.

Much of the available data suggests that increased NO_x emissions have resulted from the use of oxygenated fuels. Any increase in NO_x emissions could be detrimental in ozone nonattainment areas where exceedances have occurred during the period of the oxygenated-fuels program.

RESEARCH NEEDS

A well-designed field study with adequate statistical power should be performed at low temperatures to assess whether oxygenated fuels reduce ambient CO concentrations under such conditions. If a beneficial effect of reduction in CO is demonstrated in that field study, a carefully designed study should then investigate the effects of oxygenated fuels on emissions of NO_x, VOCs, and toxic air pollutants on winter air quality, using appropriate controls and accounting for differences in such factors as fleet population, high-emitting vehicles, inspection and maintenance programs, and local fleet characteristics. The introduction of toxic organic compounds into the air, as well as a fuel-economy penalty, should also be addressed in such a study. Also, it is equally important to perform a similar study under cold temperature conditions in an area where no oxygenated fuels are used. Results for an area using oxygenated fuels can be compared with an area not using such fuels.

FUEL ECONOMY AND ENGINE PERFORMANCE

The interagency report concludes that the fuel-economy penalty associated with the use of oxygenated fuels is approximately 2% to 3% and is related to changes in volumetric energy content. The committee agrees with these conclusions. The committee also agrees with the report's conclusion that engine performance is typically not adversely affected by the use of oxygenated fuels. The report should indicate clearly that, based on data from a wide variety of sources, fuel-economy changes are reliably predicted by the change in energy content per gallon of fuel brought about by a given change in composition.

WATER QUALITY

CRITIQUE OF INTERAGENCY DRAFT REPORT

The interagency report discussed water-quality issues arising from the use of fuel oxygenates (primarily MTBE) and their movement in the hydrologic cycle. In general, the report provides a clear and factual presentation of the current understanding of MTBE fate and transport in the environment. However, the water-quality discussion does not adequately emphasize that there is only a small amount of monitoring data available on MTBE. Such data suggest that MTBE is sometimes present in precipitation, stormwater runoff, groundwater, and drinking water.

In addition, the interagency report should clarify or revise its discussion of the following issues:

- The extent to which the results of testing for MTBE in groundwater in the Denver area might not be representative of nationwide trends.
- The discrepancy in the atmospheric degradation rates of MTBE and other oxygenates reported in the water-quality chapter versus Appendix 3 of the air-quality chapter.
- The effects of volatilization on the transport of fuel oxygenates to groundwater, especially with respect to uncertainties. The report should document the modeling approach used and the assumptions relating to MTBE concentrations in the atmosphere. The modeling approach and assumptions used to assess such effects should be documented.
- The extent to which abiotic degradation mechanisms or other

nonbiological mechanisms (e.g., chemisorption) reduce groundwater concentrations of MTBE and other alkyl ether oxygenates.

The large majority of states do not have any programs or requirements in place to monitor MTBE or other fuel oxygenates in stormwater runoff, groundwater, or drinking water. The absence of these monitoring data prevents an accurate assessment of exposure of human or aquatic biota to MTBE, possible health effects, and implementation of control measures to prevent adverse impacts.

On the basis of the small amount of monitoring data, MTBE has been detected in less than 5% of the groundwater samples analyzed, suggesting that drinking water is not currently a major exposure pathway for MTBE for much of the population.

RESEARCH NEEDS

Stormwater runoff and shallow groundwater can be contaminated with low levels of MTBE (< 20 μg/L) via precipitation or contact with small surface spills. These contamination sources should be carefully monitored to evaluate changes over time and the effect of landuse, stormwater management practices, and hydrogeologic factors on MTBE concentrations in environmental media. If EPA considerably lowers the level of its recommended health advisory concentration for MTBE, substantial concerns would arise about the potential for nonpoint sources of MTBE to adversely impact water supplies.

More needs to be known about the biodegradation of MTBE and other alkyl ether oxygenates in surface water, soil, and groundwater. Biodegradation processes, in particular, have the potential to substantially reduce the impacts of point and nonpoint source releases of MTBE and other oxygenates. Current information should be assessed to determine whether a better understanding of abiotic degradation is an important research need.

HUMAN EXPOSURE

CRITIQUE OF INTERAGENCY DRAFT REPORT

The committee finds important deficiencies in the human exposure analysis presented in the interagency draft report, which calculates cumulative exposure estimates for two hypothetical scenarios. The lifetime exposures calculated for these scenarios are 10 times higher than the maximum daily exposures estimated on the basis of empirical data as summarized in the portion of the report prepared by the Health Effects Institute (HEI). In addition to a "reasonable worst case" scenario, the report should generate a more realistic baseline exposure. Other emission products should also be considered.

The HEI-prepared portion of the interagency report provides a useful summary of MTBE exposure studies. Concentration ranges encountered in occupational and nonoccupational situations are adequately represented. Whereas the HEI report states that the data are too limited to calculate cumulative exposures for risk assessment, the committee's opinion is that the data are sufficient to bound a quantitative risk analysis and to develop a framework for conducting a comparative risk assessment of conventional and oxygenated fuels.

RESEARCH NEEDS

A quantitative framework should be established for evaluating changes in exposure from related emission products. Routine ambient monitoring of MTBE and one of its major products of photo-oxidation, tertiary-butyl formate (TBF), should be initiated in communities where MTBE is used.

Representative personal exposure monitoring of MTBE in an exposed population is needed to describe the distribution of exposures and for input into risk analyses. Such exposure monitoring should include the characterization of each individual's time-activity patterns, especially in the microenvironments where important exposures are likely to occur. The most important factors affecting personal exposure should be determined in such a study.

The relationship between fixed-site community monitoring and personal exposures to MTBE and related air pollutants should be evaluated in order to determine the information value of using community monitoring for assessing human exposure.

POTENTIAL HEALTH EFFECTS OF OXYGENATES

Critique of Interagency Draft Report

The committee is in basic agreement with the evaluation of data presented in the interagency report with respect to the metabolism, disposition, and toxicokinetics of MTBE.

The interagency report also provides a thorough review of short-term animal studies conducted on the individual oxygenates MTBE and ethanol; however, the toxicity of TBF is not addressed and should be. Based on the available data, the committee is skeptical about the need for additional toxicity studies in rats based solely on motor activity. The committee notes that even at 800 ppm the effect, if any, is 100-1,000 times predicted human exposures. It is also a reversible effect characteristic of this class of compounds, and there is no indication of neuropathology or persistent neurotoxicity following exposure to MTBE or other ethers, even after long-term exposure to high levels of MTBE.

Although no human data are available to indicate that exposure to MTBE is linked to the development of acute human disease, the committee considers it noteworthy that the available data consistently indicate that exposures to gasoline containing MTBE in occupational settings are associated with an increased rate of acute symptoms. By suggesting that "a relatively smaller proportion of persons" might have problems with exposure, the interagency report appears to ignore consistent findings in exposed workers and overlooks the possibility that typical occupational exposures may pose an overall problem, whether or not a more sensititive subpopulation exists. There are virtually no data to indicate that the reported acute health effects are confined to a sensitive subpopulation. Thus, it is the consensus of the committee that the interagency report prematurely dismisses that acute health effects might be occurring in the exposed population as a whole as opposed to occuring only in a sensitive subgroup.

The committee agrees with the interagency report that adverse reproductive and developmental effects are not expected to result from the environmental levels of MTBE to which most people would be exposed.

The report does not discuss comprehensively the long-term animal studies in their totality, i.e., a weight-of-evidence approach. While the report notes that MTBE is a multispecies, multisite, and multisex animal carcinogen, it fails to make note of certain inconsistencies in the data. Because of inconsistencies and unresolved questions with regard to the animal carcinogenesis studies, "cancer potency estimates" of MTBE as proposed in the report should be considered cautiously. The committee feels that the male rat kidney-tumor data probably should not be used for this purpose in light of the new information on its probable causation, i.e., α_{2u}-globulin nephropathy, which is thought to be unique to the male rat and not relevant to humans. The use of the lymphoma and leukemia data should also be questioned until such time as a thorough review of this study, including an objective third-party review of the pathology, is accomplished. The most reliable data presently available for risk-assessment purposes are on the induction of benign liver tumors in female mice exposed via inhalation to 8,000-ppm MTBE.

RESEARCH NEEDS

Additional research on the toxicokinetics of MTBE and other oxygenates would be useful as a basis for extrapolating health-effects data from animals to humans and for identifying suitable biologic markers of exposure for use in any future epidemiological studies.

The committee agrees with the report that additional short-term animal studies on MTBE that determine blood levels in addition to evaluating central nervous system (CNS) function could be useful.

The committee believes there should be better coordination between clinical observations, epidemiologic studies, and exposure-chamber experiments regarding research on acute human health effects and exposure to oxygenated fuels. Improved clinical characterization of symptoms reported to be attributed to exposures from MTBE in gasoline is also needed and should take into account the actual types of symptoms experienced by individuals, the settings in which the symptoms occur (e.g., refueling versus driving) and the duration of symptoms. Developing objective measures of outcomes representing the key symptoms initially defined by the Alaska studies is desirable where possible. Research in the development of improved methodology in the field of acute health effects, including validation and reliability of instruments used to measure symptoms without physical findings, should be supported. One or more analytical epidemiologic studies examining the association between MTBE exposure and acute health effects should be conducted. These studies should include individual-level quantitative exposure assessments, outcome assessments through questionnaires whose reliability and validity have been established with pilot data or through objective measurements, and high-quality data on potential confounders, including demographics, weather conditions, concurrent exposures, and automobile characteristics. (See Chapter 5 for study details.) Due to difficulties (e.g., high cost, long follow-up time, large required sample size) in conducting an analytical epidemiologic study to investigate the relationship between cancer and other chronic diseases and long-term exposure to MTBE, attention should be turned to ecologic designs, despite their widely acknowledged deficiencies. However, the committee does not recommend that ecological studies be undertaken at the present time, but rather that missing routine environmental-monitoring data on air and water begin to be collected, so that such studies can be conducted in the future.

POTENTIAL HEALTH EFFECTS OF OTHER POLLUTANTS

CRITIQUE OF INTERAGENCY DRAFT REPORT

The committee agrees with the interagency report that the data are too limited to evaluate the effectiveness of the winter oxygenated-fuel program in lowering CO exposures to a level that would not affect cardiovascular morbidity and mortality.

RESEARCH NEEDS

Epidemiologic studies are needed to evaluate the effectiveness of the winter oxygenated-fuel program in lowering CO exposures to a level that would reduce carboxyhemoglobin levels and risk of exacerbation of cardiovascular disease. Epidemiologic studies are also needed to determine the exposures to aldehydes and air toxics, produced from the combustion of oxygenated fuels and from conventional gasoline, that would cause irritant and other short-term health effects.

RISK ASSESSMENT

CRITIQUE OF INTERAGENCY DRAFT REPORT

With respect to the noncarcinogenic effects of MTBE resulting from long-term, continuous inhalation, the interagency report's use of EPA's reference concentration (RfC) of 3 mg/m^3 (0.8 ppm) seems appropriate. This value should not, however, be used to judge the safety of acute exposures until the possible association between MTBE exposures and acute symptoms is resolved. With respect to carcinogenic risks, all estimates presented in Table 7.3 of the interagency report that are based on rat kidney tumors and lymphomas and leukemias should be eliminated until the investigations recommended above are completed (see Chapter 5). Moreover, maximum likelihood estimates (MLEs) are not reliable and should not be presented. The failure of the interagency report to make any comparison of the risks of MTBE-containing fuels and nonoxygenated fuels is a serious deficiency and should be corrected. All available data should be considered, including various surrogate measures of exposure, to compare risks and then the degree of uncertainty associated with those comparisons should be stated. The presentation of cancer risk estimates for MTBE in isolation from all other risks has little value for risk managers. The assessment contained in the HEI report comes close to the type of evaluation the committee thinks is useful for decision-making. The overall conclusion of the HEI report, that MTBE-containing fuels do not pose health risks substantially different from those associated with nonoxygenated fuels, seems reasonably well supported but requires additional quantitative documentation.

RESEARCH NEEDS

The National Research Council committee suggests that the portion of the report that was prepared by HEI be used as the framework and database (with appropriate additions and reinterpretations as described by the National Research Council committee) for a comprehensive government risk assessment. However, greater effort should be made to provide some indication of the magnitudes of the health risks that are said to be increased and decreased (relative to conventional fuels) by the use of oxygenated fuels.

COSTS AND BENEFITS

The interagency report does not present an assessment of the costs and benefits of the oxygenated-fuels program. Although the committee was not charged to conduct such an assessment, it concludes that the interagency report should address this issue. Despite uncertainties in estimating program-related costs and even greater uncertainties in estimating benefits, full evaluation of the winter oxygenated-fuel program requires that the interagency report address and document program costs and benefits at least at a broad level.

Toxicological and Performance Aspects of Oxygenated Motor Vehicle Fuels

I

INTRODUCTION

The Clean Air Act Amendments of 1990 mandated the use of fuels with higher oxygen content (at least 2.7% by weight) in several areas of the country that are in nonattainment with the National Ambient Air Quality Standards (NAAQS) for carbon monoxide (CO). The use of such oxygenated fuels is intended to reduce CO emissions from motor vehicles during the winter months, when lower temperatures tend to cause vehicles to emit more CO. Increased ambient concentrations of CO are a particular concern for people with cardiovascular disease. In this report, "oxygenated fuels" is intended to refer only to the fuels used within the winter oxygenated-fuels program.

Methyl tertiary-butyl ether (MTBE) has become the most widely used motor vehicle fuel oxygenate in the United States. Typically, MTBE-oxygenated gasoline contains approximately 15% MTBE by volume. Other compounds used as oxygenates in gasoline include ethanol (which is the dominant oxygenate in some areas), tertiary-butyl alcohol, ethyl-tertiary-butyl ether, and tertiary-amyl-methyl ether.

Since the 1970s, MTBE has been added to motor-vehicle fuels as an octane booster, in concentrations of less than 1% by volume in regular gasoline and 2% to 9% by volume in premium gasoline (USEPA, 1993). The dilution of gasoline with MTBE reduces certain other organic compounds (e.g., benzene) in gasoline. The intended result for motor vehicle emissions is decreased amounts of CO and of some toxic air pollutants. However, increased amounts of MTBE and byproducts, such as formaldehyde, are also emitted.

Concurrently with the start of the federal oxygenated-gasoline program in 1992, MTBE has been implicated in complaints of headaches, coughs, and nausea, principally in Alaska, but also in Montana, New Jersey, and Wisconsin. There have also been anecdotal reports of reduced fuel economy in some locations and questions about engine performance. Additional concerns have been raised about the detection of low levels of MTBE in some samples of groundwater. Also, use of oxygenated fuel has increased costs of gasoline at the pump by several cents per gallon.

Due to the heightened public concern over the potential health effects of MTBE-oxygenated fuels, scientific studies costing more than $2 million have been conducted by EPA and others, in order to investigate the reported symptoms. These studies included both experimental human studies with pure MTBE and epidemiologic studies of MTBE mixed with gasoline. As of this date, the reported symptoms have not been explained or reproduced experimentally. However, due to the widespread pattern of exposure, the possibility of susceptible subpopulations, and reports of other noncancer effects, EPA has also concluded that more research and testing are required to better understand the comparative risks of different fuels.

In its continuing efforts to evaluate the toxicological effects and benefits of oxygenated gasoline, EPA asked the National Research Council to conduct a brief study to review a draft interagency report from the federal government, prepared under the direction of the Office of Science and Technology Policy (OSTP) through the Committee on Environment and Natural Resources (CENR) of the president's National Science and Technology Council (NSTC).

INTERAGENCY REPORT

The Interagency Oxygenated Fuels Assessment draft report was prepared by a working group comprising technical and scientific experts from several federal agencies, as well as representatives from state government, industry, and environmental groups (Interagency Report, 1996). The draft interagency report was provided to the committee on March 15, 1996. (The preface and executive summary of the interagency report are presented in Appendix A.) The interagency report discusses MTBE to a much greater degree than other oxygenates.

The scientific literature reviewed by the authors included published, peer-reviewed literature; unpublished reports from a number of sources, including industry, government agencies, and scientists; and personal communications. The potential health effects of oxygenated gasoline were evaluated in three separate documents, which made up the health effects section of the interagency report. The three documents are the following:

- A report from the Health Effects Institute (HEI) Oxygenates Evaluation Committee, *The Potential Health Effects of Oxygenates Added to Gasoline: A Review of the Current Literature* (February 1996), referred to as the HEI report;
- A report from CENR and the Interagency Oxygenated Fuels Assessment Steering Committee, *Interagency Assessment of Potential Health Risks Associated with Oxygenated Gasoline* (February 1996), referred to as the OSTP report; and

• A memorandum from the Centers for Disease Control and Prevention (CDC) signed by Richard J. Jackson and directed to the Interagency Oxygenated Fuels Assessment Steering Committee, dated March 12, 1996 (referred to as the CDC white paper), which compared and contrasted the other two reports.

The interagency report also included documents addressing air quality (Howard et al., 1996), water quality (Zogorski et al., 1996), and fuel economy and engine performance.

Each of the chapters in the interagency report underwent extensive external peer review prior to the submission of the entire report for review by the National Research Council. The draft interagency report will be revised in response to the findings and recommendations of the National Research Council (Robert Watson, OSTP, April 1, 1996, personal communication).

CHARGE TO THE NATIONAL RESEARCH COUNCIL COMMITTEE

In response to EPA's request, the National Research Council convened a multidisciplinary committee—the Committee on Toxicological and Performance Aspects of Oxygenated Motor Vehicle Fuels. The committee was charged with the following tasks: (1) review the draft interagency report on the potential toxicological effects of MTBE and other oxygenates and compare such effects and multimedia exposures with those of conventional gasoline; (2) review the draft interagency report on the impacts of MTBE and other oxygenates on vehicle emissions, air pollution, fuel economy, and engine performance and compare such impacts with those of conventional gasoline; and (3) identify priorities for research to fill data gaps.

The committee was asked to focus on the interagency report,

which dealt exclusively with the winter oxygenated-fuels program. The interagency report did not specifically examine the reformulated gasoline program, which is intended to reduce motor-vehicle emissions that lead to increased ozone concentrations in the lower atmosphere. Thus, the committee did not address the reformulated-fuels program. Also, the committee was not charged to assess the costs and benefits of the oxygenated-fuels program.

COMMITTEE APPROACH

The committee members were provided with the six draft documents described previously prior to their only meeting. A full and complete critique of scientific credibility, comprehensiveness, and internal consistency of the data was not possible within the short time-frame of this study. In general, the committee responded to its charge by first summarizing critical aspects of the interagency report upon which its critique was based. The committee's critique was followed by recommendations for future research.

STRUCTURE OF THE REPORT

The results of the committee's deliberations are found in the chapters of this report. Chapter 2 is a critique of sections of the interagency report related to air quality, fuel economy, and engine performance. Chapter 3 and Chapter 4 critique water-quality and exposure-assessment discussions of the interagency report, respectively. In Chapter 5, the report evaluates the federal government's assessment of the potential health effects of exposure to oxygenates, as presented in three reports (the HEI report, OSTP report, and the CDC white paper) contained in the interagency report. Chapter 6 evaluates the potential health effects of other pollutants resulting from the combustion of oxygenated fuels and conventional gasoline. Chapter 7, the final chapter, provides the committee critique of the approaches to the assessment of human health risk adopted in the HEI and OSTP reports. It should be noted that each chapter provides recommendations for research efforts.

2

Air Quality, Fuel Economy, and Engine Peformance

A draft interagency report has been prepared that assesses the impact of winter use of motor-vehicle fuels oxygenated with compounds, such as methyl tertiary-butyl ether (MTBE), on air quality, fuel economy, engine performance, water quality, and potential health effects. This chapter is a critique of the section on air-quality benefits and the section on fuel-economy and engine-performance issues of the interagency report. It concentrates primarily on major deficiencies and omissions in those sections of the interagency report.

The committee's evaluation should not be construed as a criticism of the interagency's discussion of air quality, fuel economy, and engine performance. More precisely, this evaluation is a critical review of the quality and quantity of data available from which to draw conclusions regarding the effectiveness of oxygenated fuels in reducing ambient concentrations of carbon monoxide during winter.

Under the Clean Air Act Amendments of 1990, the U.S. Environmental Protection Agency (EPA) mandates the use of oxygenated fuels in winter in areas exceeding the National Ambient Air

Quality Standard (NAAQS) for carbon monoxide (CO). The interagency report was prompted by health effects claimed to result from use of oxygenated fuel. Most of these claims were concentrated in the colder areas, which exceeded the CO NAAQS.

Table 3 in the portion of the interagency report titled "The Potential Health Effects of Oxygenates added to Gasoline - A Review of the Current Literature" lists the areas participating in the oxygenated-fuels program in 1994-1995; these areas include cities with a broad range of average winter temperatures, e.g., Minneapolis and San Diego. For reasons explained later, there is an inverse relationship between ambient temperature and the amount of tailpipe CO emissions. Consequently, it is important that the term "winter temperature" be defined. The interagency report does not do so. A map, or similar representation, giving winter temperature ranges and mean temperature for the participating areas is essential in understanding and evaluating the data.

A major deficiency in the interagency report is that it does not give the reader information that will permit the needed assessment of the interrelationship between exhaust emissions measured during the Federal Test Procedure (FTP) and exhaust emissions produced during on-road use, especially in winter. FTP is a standard procedure for measuring vehicle exhaust emissions at 75°F (actually between 68°F and 86°F). As explained below, there are major differences between emissions at 75°F and at low termperatures.

To understand how the winter use of oxygenated fuels in vehicles can affect air quality, several factors must be taken into account in a general way:

(a) The effect of oxygenated fuels on tailpipe emissions—i.e., CO, volatile organic compounds (VOCs), and oxides of nitrogen (NO_x)—and emissions due to evaporation and running losses of fuel from the tank and fuel lines. It is important to account for how such effects differ among individual vehicles using different emission-control techniques, having different maintenance histories, and operating at typical winter running conditions. A brief summary of the factors influencing tailpipe emissions is provided later in this chapter.

(b) The inventory or population of different vehicles, including normal and high emitters in the existing fleet.

(c) Particular meteorological conditions that might affect dispersion and transformation of the emitted pollutants.

It is by no means easy to integrate these factors to assess the effect of the implementation of any new regulation or technology. Ultimately, measurements of ambient-air quality provide the best assessment. Thus, any attempt to transform results from individual-vehicle operation into resulting air-quality benefits must be made with extreme caution.

Tailpipe emissions represent the difference between formation of pollutants in the engine combustion process and destruction (through oxidation or reduction) of these pollutants in the exhaust system. The most important factor influencing the formation of pollutants is air-to-fuel (A/F) ratio; the most important device used in destroying them is the catalytic converter.

Figure 2.1 illustrates the general relationship between A/F ratio and pollutant production. Although fuel constituents other than oxygen (O_2) are known to have an impact on emissions when oxygenated fuels are used, the primary factor in decreasing CO concentrations is increasing the A/F ratio through the O_2 brought in with the fuel (enleanment). To illustrate this fact, Figure 2.2, for methanol, indicates that, if the same stoichiometry between normal fuel and oxygenated fuel is maintained, there is little change in CO emissions. In addition, A/F ratio has a substantial effect on power, fuel economy, cold starting, and drivability.

Figure 2.3 illustrates the importance of catalyst temperature on destruction of CO and VOCs. If the A/F ratio is closely main-

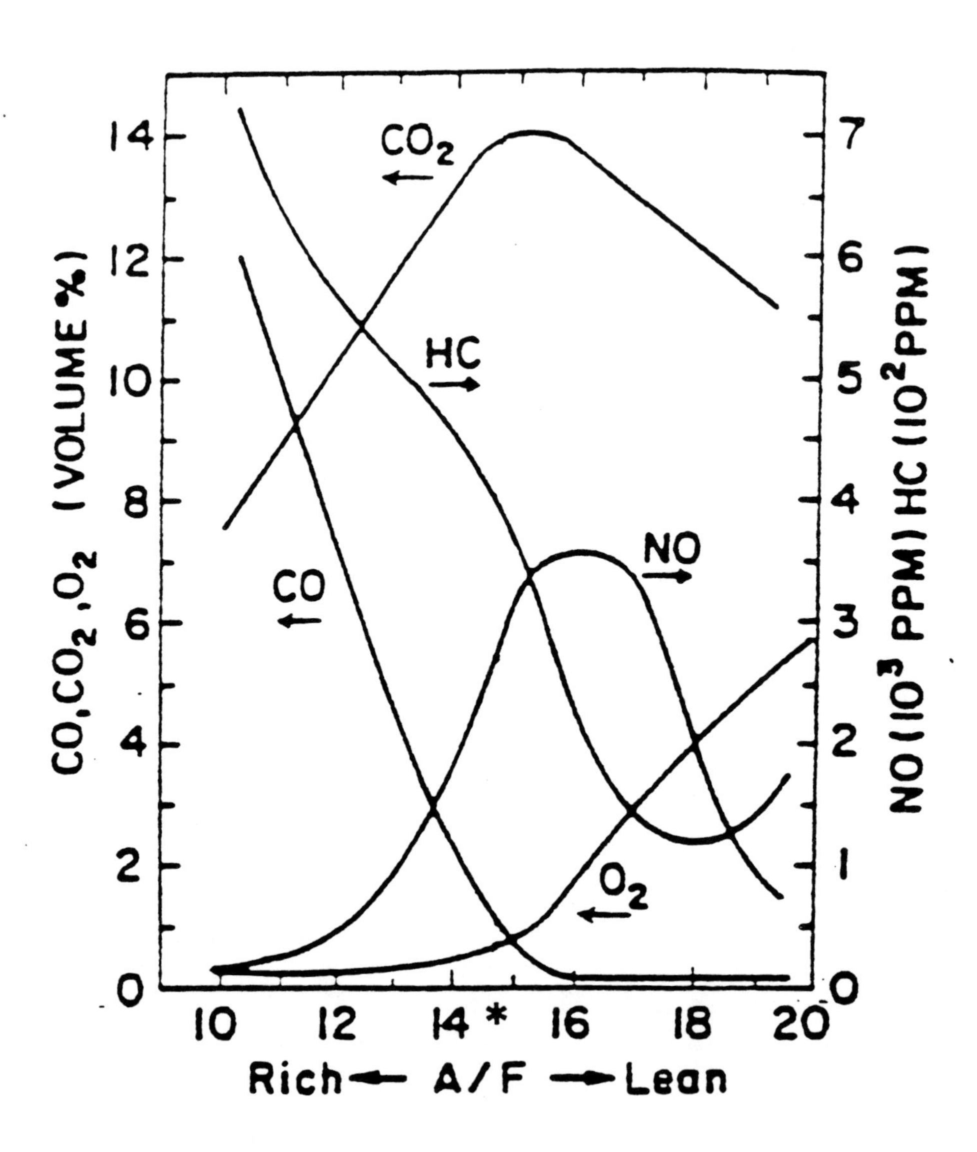

Figure 2.1. Effect of A/F ratio on gasoline-engine pollutant emissions. * represents the stoichcometric A/F ratio. Adapted from Kummer, 1980.

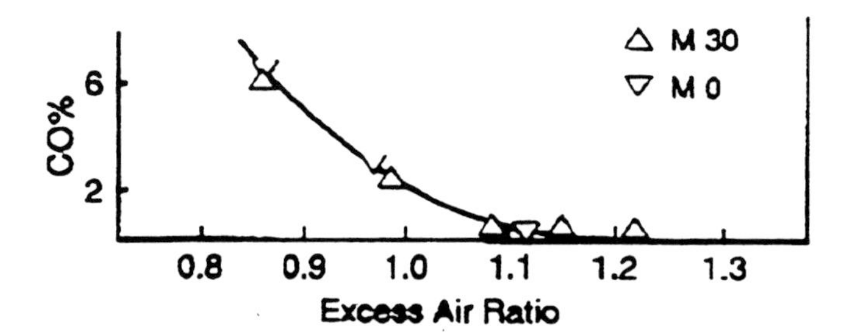

Figure 2.2 Relationship between CO emissions and excess air ratio for M30 (gasoline and methanol (30% vol)) and M0 (gasoline). Excess air ratio is the amount of O_2 (from air and fuel) divided by the stoichiometric amount of O_2 needed to completely combust the fuel to CO_2 and water. An air-cooled single-cylinder engine was used (compression ratio 6.8, 1500 rpm, ¼ load, and spark timing-maximum best timing). Adapted from Hirao and Pefley, 1988.

tained around a stoichiometric level (Figure 2.1), modern catalysts will destroy CO, VOCs, and NO_x very efficiently.

For good drivability, different values of A/F ratios are needed during vehicle operation. For starting the vehicle, extra fuel must be supplied because not all the fuel vaporizes under those conditions. The extra fuel needed decreases as the engine warms up. Rapid accelerations may require momentary enrichment because A/F ratio affects power. If on-road accelerations are greater than those used in the standard test cycle (Figure 2.4), actual emissions may differ from emissions measured during the FTP.

If the A/F ratio in the cylinder is too lean, or too rich, the engine may hesitate or stall, i.e., drivability will be poor. Consequently, for both emission and drivability reasons, an A/F-ratio control system is generally used. A/F-ratio control can be generally divided into two classifications: open-loop and closed-loop. Generally, in open-loop control, A/F ratios are predetermined (typically

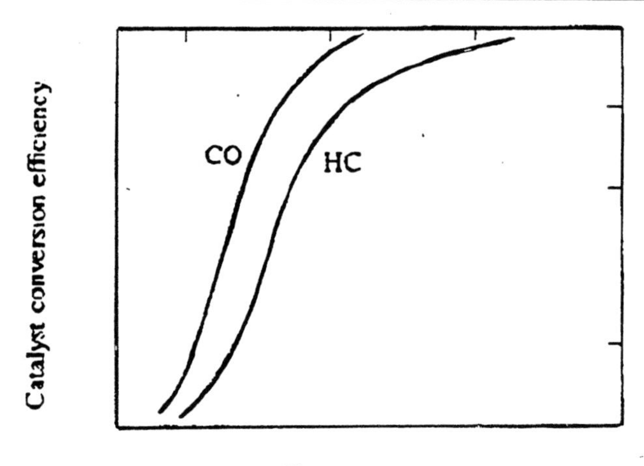

Figure 2.3. Illustrative relationship between motor-vehicle catalyst temperature and destruction efficiency of CO and hydrocarbon (HC) emissions. Adapted from Heywood, 1988.

stoichiometric or richer) but changed by ambient and operating conditions.

In closed-loop control (Figure 2.5), the A/F ratio is automatically adjusted to achieve a given goal, in this case maintaining the stoichiometric mixture needed to destroy all three pollutants. If the closed loop adapts or adjusts to changing conditions, it is called adaptive closed-loop control.

Carbureted engines use open-loop control. However, almost all current engines, using fuel injection and computer control, operate initially as open-loop but change to closed-loop as the intake manifold, O_2 sensor (increasingly electrically heated), and catalytic converter warm up. While not shown in Figure 2.5, a second O_2 sensor, located downstream of the catalytic converter, is increasingly being added to achieve more precise control.

Because of differences with vehicle age, model, and manufacturer, a single value for the duration of open-loop operation cannot be presented. For FTP conditions at 75°F and computer-controlled

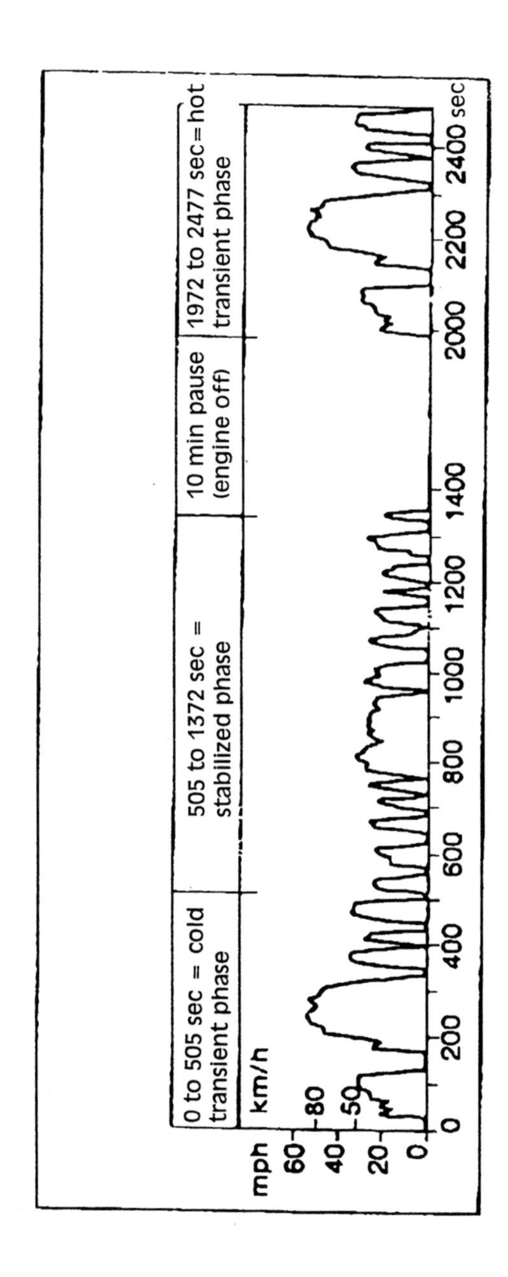

Figure 2.4. The Federal Test Procedure driving cycle. The cycle length is 11.115 miles; cycle duration is 1877 sec with a 600 sec pause; average speed is 34.1 km/hr (21.2 mile/hr); and the maximum speed is 91.2 km/hr (56.7 mile/hr). The procedure is conducted at a temperature of 75°F. Adapted from: Bosch, 1986.

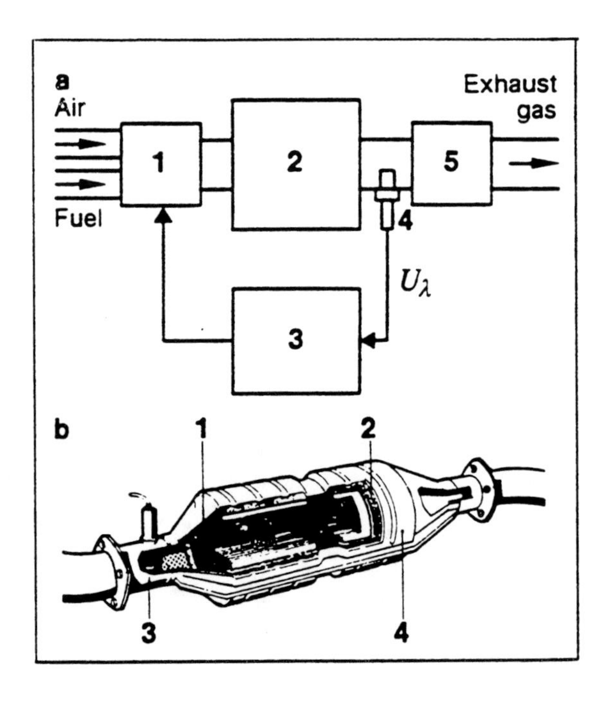

Figure 2.5 Exhaust-gas emission control with the three-way catalyst and the O_2 sensor. a) Functional diagram: 1, fuel-metering system; 2, engine; 3, control unit; 4, O_2 sensor; 5, three-way catalyst; $U\lambda$, sensor voltage. b) Three-way catalyst: 1, ceramic monolith; 2, wire mesh; 3, O_2 sensor; 4, housing. Adapted from Bosch, 1986.

operation, the open-loop time is 30-80 seconds. Depending upon the control technique, engine design, and catalytic-converter location, catalyst "light-off" will occur 0 - 80 seconds after conversion to closed-loop control. For a cold winter startup, open-loop operation will be longer—possibly 2 or 3 times longer.

Given the wide range of control techniques used and maintenance histories of vehicles, it is difficult to generalize about the effect of using oxygenated fuels on CO emissions. However, during closed-loop operation with perfect control there should be little change in CO emissions; the control sensor should maintain the same stoichiometry and therefore the same CO emissions, irrespective of fuel.

For open loop operation, the situation is more complicated. In general, rich mixtures are used during open loop operation in accordance with a pre-determined schedule of A/F, spark advance, etc, which is modified by ambient and engine conditions. Adaptive control is defined as a change in the values in this schedule made automatically in response to a change in fuel characteristics. If an oxygenated fuel is used and adaptive control is not used, enleanment will occur resulting in reductions of CO emissions. In addition, the potential for NO_x increases (see Figures 2.1 and 2.2), and vehicle driveability is likely to decrease.

Use of adaptive control began approximately a decade ago. Although the first systems were rather crude in sensing when oxygenated fuels were used, current adaptive control systems adequately sense when oxygenated fuels are being used and correct the open-loop schedule for the effects of oxygen in the fuel. Thus for very recent and future vehicles operating in open loop and using oxygenated fuels, enleanment and driveability problems would be minimal. For current and future vehicles, in both open loop and closed loop operation, only small changes in CO and NO_x emissions can be expected when using oxygenated fuels.

The interagency report does not provide a plausible method for adjusting FTP data collected at the standard temperature of 75 °F to represent data that would be observed at winter temperatures. Although data for temperatures below 75 °F are presented, it is not clear whether the behavior of vehicles under such tests would adequately represent typical winter conditions. The interagency report should discuss this deficiency and its implications on the report's conclusions.

An alternative procedure for assessing the effect of oxygenated fuels on ambient CO is to demonstrate that ambient levels of CO have been reduced by the use of oxygenated fuel. This case is not made in the interagency report, perhaps because it is difficult to separate the reduction due to oxygenated fuels from the general reduction in CO due to reduced production by the general fleet. Figure 1.1 of the air-quality benefits section of the interagency report presents such a plot for Denver covering the period 1980 - 1995. While the annual variation in CO levels has decreased, the trend of average winter CO level shows little or no change since winter oxygenated fuels were introduced.

The interagency report states that EPA's MOBILE 5a model has predicted decreases as large as 29% in vehicle CO emissions resulting from the use of oxygenated fuels. Because 90% or more of the ambient CO in urban areas results from emissions from mobile sources, a predicted decrease of 29% in mobile-source CO emissions presumably should be observable in the ambient data. This has not been the case.

Another important deficiency of the interagency report is the failure to emphasize the fact that, from a mass-emission standpoint and consequent effect on ambient CO levels, high emitters are the major contributors to ambient CO concentrations. High emitters are vehicles whose emissions are inordinately large as a result of older control techniques, misadjustment, lack of maintenance, etc. Their emissions can be orders of magnitude larger than those from current, well-maintained vehicles. It follows that the same percentage reduction in emissions from high emitters will produce orders of magnitude greater mass-emission reductions than will the same percentage reduction from low emitters.

In 1983, a contractor reported to the California Air Resources Board (CARB) that 12% of a fleet of vehicles recruited by CARB for its in-use surveillance program was responsible for 47% of the total CO emissions as measured by the 75°F FTP (Wayne and Horie, 1983). The contractor recommended: "Since outliers [high emitters] contribute a large fraction of the total emissions, it is more important to know their contribution accurately than to know accurately the relatively minor contributions of low-emitting vehicles." This same principle applies to any study of CO emissions from the in-use fleet. Since that report, a number of additional studies have reported the same findings—at least half the CO emissions from warm vehicles under stable operating conditions are produced by less than 10% of the fleet (Bishop and Stedman, 1990; Stephens and Cadle, 1991; Lawson, 1993; Zhang et al, 1995; Shepard et al, 1995). Some of these high-emitting vehicles produce more than 0.5 pound (about 227g) of CO per mile (Lawson et al, 1996).

Because the CO emissions of the in-use fleet are highly skewed, it is important to understand which control strategy will produce the largest reduction in CO emissions at the lowest cost to society (Beaton et al, 1995). For example, a relatively large percentage decrease in emissions from well-maintained vehicles with modern control technology will result in only a minor total reduction in ambient CO emissions.

Finally, another important issue that needs additional discussion and clarification is results obtained with EPA's MOBILE 5a model. EPA designed the MOBILE model specifically for use by states in preparing emission inventories required under the Clean Air Act. MOBILE 5a is the model version currently in use. The model is used to estimate the current fleet's tailpipe and nontailpipe emissions, the effectiveness of vehicle inspection and maintenance programs, and the effect of oxygenated and reformulated fuels on vehicle emissions. It is also used to provide projections of the vehicle fleet's emissions many years into the future, and output from this model is used to provide emissions data that go into our nation's emission inventories. Predictions from this model are used in the interagency report but, in spite of differences between projected and real-world data, no assessment of the accuracy of this model is presented.

FTP EMISSIONS DATA FOR INDIVIDUAL VEHICLES

The available CO data in the interagency report for individual vehicles used in the study were collected largely as standard FTP 75°F cycle emissions. The report covered data from emission-control systems that ranged from early technology to recent technology. There were some data comparing high emitters and normal emitters using the standard 75°F FTP cycle. The report presents some data using the FTP cycle at 35°F and 50°F. The data for 35°F and 50°F test conditions were for relatively clean vehicles as compared to high emitters.

The primary objective of the oxygenated-fuels program is the reduction of winter CO emissions in nonattainment areas. However, CO emissions data are lacking for the temperatures less than 20°F for both FTP and on-road conditions. Such conditions are typical of winter conditions in many regions of the United States. (The normal daily minimum temperature in Fairbanks, Alaska, in January is about -22°F.) When operating at temperatures less than 20°F, vehicles require much richer startup mixtures for much of the open-loop operation, may have reduced catalyst temperatures, and warm up slowly depending upon driving conditions, such as stop-and-go traffic, freeway driving at high speed, and congested freeway driving. Such data are extremely critical in judging the benefits of the winter oxygenated-fuels program, especially because some of the data indicate tendencies for increased CO emissions with oxygenated fuels. High-emitting vehicles also tend to be highly variable in their measured emissions (Knepper et al., 1993). The interagency report should provide a single summary table that lists the results of all the dynamometer-based studies, the testing conditions and types of vehicles tested (normal versus high emitters), with changes in observed pollutant emissions.

ON-ROAD EMISSIONS DATA FROM VEHICLE FLEETS

Field studies of emissions from the on-road motor vehicle fleets are necessary to quantify the real-world emissions coming from the tailpipes of vehicles. There are two types of studies that can be used to infer the levels of CO emissions from in-use vehicles and the effect of oxygenated fuels in the winter. One approach uses remote sensing data and the other uses data acquired in tunnel studies, in which vehicle emissions are measured as the fleet is driven through a tunnel.

The interagency report discusses results from remote-sensing studies from the Denver area and North Carolina. In the first Denver study, Bishop and Stedman (1989) reported a $6 \pm 2.5\%$ reduction in CO attributable to the use of 1.5-wt% oxygenated fuel. In the second Denver study, Bishop and Stedman (1990) reported a $16 \pm 3\%$ decrease in CO emissions from the use of 2.0-wt% oxygenated fuel. Another study, by PRC (1992), reported that the percentage reduction of CO emissions was nearly the same for all vehicles and that most of the CO emissions attributed to oxygenated-fuel use were from the highest-emitting vehicles.

However, two additional remote-sensing studies performed in

North Carolina by Georgia Tech (Rhudy et al. 1995; Rodgers, 1996 [personal communication]) have shown no reduction in CO emissions that could be attributed to the use of oxygenated fuels. For all these studies, remote sensing devices measure the CO-to-CO_2 ratio in the plume of vehicles that pass by the sensors. Because CO and CO_2 emissions increase dramatically at lower-temperature winter conditions, it must be demonstrated that the ratio of these two pollutants from the in-use fleet is not adversely affected at cold temperatures for different fuel compositions in order for this type of experiment to be reliable. A decrease in measured CO-to-CO_2 ratios does not necessarily represent a decrease in CO emissions; it also could result from an increase in CO_2 emissions from higher fuel consumption (as explained later in this chapter). Finally, because different groups have not obtained consistent experimental results, at this time the committee cannot conclude that the remote-sensing results have conclusively demonstrated a beneficial effect from oxygenated fuels.

In the second category of study, Kirchstetter et al. (1996) report a 21% reduction in the fleet's CO emissions in the Caldecott tunnel in the San Francisco area. However, the observed differences in CO emissions occurring before and during oxygenated-fuel use might not be entirely attributable to differences in fuel composition. Other factors, such as differences in typical speeds of the vehicles, might have affected the observed CO reduction. In order to obtain more conclusive results, a set of measurements should have been made after the oxygenated-fuel season ended. Also, it is important to note that measurements were taken at temperatures higher than temperatures characteristic of winter in many areas. Cold-start conditions are not measured in a tunnel; for the current fleet, such conditions provide the greatest potential emission reductions attributable to oxygenated fuels. Finally, a 21% reduction in CO would be expected to result in a marked decrease in observed ambient measurements of CO. Such changes have not been observed.

In each of these studies, a suitable experimental control group or control conditions have not been used. In the case of in-use fleet-emission experiments, precise documentation of the number of high-emitting vehicles and their actual emissions also is necessary because of their large influence on overall fleet-emission characteristics.

LOW-TEMPERATURE VEHICLE-EMISSIONS STUDIES

As ambient temperatures decrease, the amount of time that a vehicle runs in open-loop conditions increases substantially as does the catalytic converter warmup time. The length of time that vehicles are actually in open loop operation is not well documented, however. Because increases in CO emissions at low temperatures are very large, the interagency report should discuss this effect. Also, that report should compare such increases in CO emissions with reductions in CO attributable to oxygenated fuels.

The dynamometer data cited in the report suggest an extremely small reduction in CO emissions when vehicles are driven in FTP-like conditions at low temperatures (less than 35°F). Studies performed by EPA and others at low temperatures should be included in the air-quality chapter. In the low-temperature regions in the United States where oxygenated fuels are required, it is clear that additional data are needed to understand whether the use of oxygenated fuels actually decreases CO emissions. The committee recommends that the air-quality chapter of the interagency report contain a summary table which presents the results of all low-temperature dynamometer studies.

AMBIENT STUDIES

The interagency report cites studies of ambient measurements by the state of California (Dolislager et al., 1993, 1996), Centers for Disease Control and Prevention (CDC) (Mannino and Etzel, 1996) and the University of Colorado at Denver (Anderson et al., 1993, 1994) of ambient CO data from western states and additional studies from Alaska (Heil, 1993) and North Carolina (Vogt, 1994, Vogt et al., 1994). Only one of those studies has appeared in peer-reviewed literature; the rest are only in the gray literature. The peer-reviewed paper, Mannino and Etzel (1996) however, has a number of serious shortcomings. It states that oxygenated fuels reduce ambient CO concentrations, but it also reports that "the largest [CO] decreases were at two sites that were not using oxygenated fuels.... At some monitors, however, CO levels increased slightly." The report also states that "these decrements were small in terms of absolute concentrations" The authors did not explain why these areas had increases or only small decreases in ambient CO concentrations. During the period of the CDC study, the frequency of occurrence of CO exceedances in oxygenated-fuel air-monitoring locations decreased from 66% to 28%, while the percentage of nonoxygenated-fuel air-monitoring locations with CO exceedances decreased from 52% to 11%. There was a larger decrease of CO exceedances in conventional-fuel areas than in areas using oxygenated gasolines during the CDC study period. Most important, however, the CDC report did not account for the influence of motor-vehicle inspection and maintenance (I/M) programs in their analysis. In addition to oxygenated fuels, CO nonattainment areas are required to have I/M programs in place. According to EPA's MOBILE model, CO reductions resulting from I/M programs are substantial. However, as discussed previously the accuracy of the MOBILE model is uncertain. It is possible that I/M programs were responsible for the larger CO decreases in the oxygenated-fuel regions than in the conventional-fuel (and non-I/M-program) regions. For example, California's mobile-source emissions model predicts a decrease of 15% in the fleet's CO emissions resulting from that state's I/M program (California I/M Review Committee, 1993).

Although the calculations appear to be thorough, Dolislager (1993, 1996) did not perform an analysis of ambient concentrations and ratios in the three separate periods (early fall, winter, and early spring) for the several years before oxygenated fuels were used in California to determine whether CO, VOC, and NO_x relationships were similar for those three seasons when oxygenated fuels were not used. A similar comparison for the three seasons during years when oxygenated fuels were used would allow an oxygenated-fuels effect to be adequately evaluated. Also, Dolislager calculated changes in CO-to-NO_x ratios to infer changes in CO emissions due to oxygenated fuels by assuming that oxygenated fuels do not substantially increase NO_x emissions. The data from the interagency report show that oxygenated fuels do increase NO_x emissions by as much as 2% per weight % oxygen. Thus, the actual CO reductions resulting from the use of oxygenated fuels would be lower than the 6-10% reported. The California analysis reports a 6-10% decrease of ambient CO attributable to the use of oxygenated fuels, and it should be noted that this decrease has been observed in areas having very mild winters. As is noted in other parts of this report, low-temperature dynamometer data show extremely small (only 1.8-1.9%) decreases in CO emissions per weight % of oxygen (in some cases, CO emission increases have been reported).

Two separate reports have been written by Keislar et al. (1995, 1996) summarizing an ambient study in Provo, Utah, designed to assess the effect of oxygenated fuels on ambient CO concentrations. In the first report, the authors calculated an upper limit of 9-10% decrease in CO but at a confidence level of only 80%. In the second paper, the authors cited a benefit of $15 \pm 20\%$.

Several additional ambient-effects studies cited in the interagency report show no observable effect on ambient CO from the use of oxygenated fuels. The committee feels that the wording in the conclusions section of the interagency report should explicitly state that several locations where oxygenated fuels have been used have shown no effect of the fuels on reduction of ambient CO concentrations and that in some cases CO concentrations have actually increased. The committee recommends that the air-quality chapter contain a summary table that presents the ambient conditions, reported uncertainties, and results reported in all the ambient studies.

From the data cited in the interagency report, the effects of oxygenated fuel on ambient levels of CO are small at best; in some locations, increases in ambient CO have actually occurred (Mannino and Etzel, 1996). In some locations having moderate climates, (e.g., California), a 5-10% effect has been reported (Dolislager, 1993, 1996); in another moderate climate location (e.g., North Carolina), no effect was observed (Vogt, 1994; Vogt et al., 1994; Cornelius, 1995). Only one of these studies has been subjected to the peer-review process, and that report, as discussed in the committee's report, suffers deficiencies. Few or no data from low-temperature winter locations exist: only one report from Alaska. The major problem is a lack of thorough, statistically defensible analysis of ambient data where such confounding features as the lack of a control population for comparison, fleet turnover, occurrence of high emitters, I/M programs, and local economy and fleet population are properly accounted for.

DISCREPANCIES BETWEEN MODEL RESULTS AND OBSERVATIONS

The interagency report also suggests that there are major discrepancies between real-world observations and the results of the MOBILE 5a model. Since 1987, there have been a number of independent studies by researchers throughout the country that document inadequacies in EPA's MOBILE model (Ingalls et al., 1989; Pierson et al., 1990). The observed CO reductions attributed to oxygenated fuels are much less than predicted by the MOBILE model. The model's predictions for CO reductions, as cited in the report, are as large as 29%. No ambient study reports reductions in CO anywhere close to this value.

It should be noted that variations in factors including meteorological conditions, monitor location, engine operating conditions, vehicle fleet composition, travel activity, and variability in measurements can make it difficult to detect statistically significant changes in ambient concentrations. However, when ambient studies experience difficulties in detecting the effects of emission control programs, it suggests that the effects are not major and that additional analysis would be needed to isolate the effect.

The interagency report should include an assessment of the utility of the MOBILE 5a model for prediction of winter emissions from oxygenated fuels and discuss possible reasons for differences between observations and modeled predictions.

COPOLLUTANT EFFECTS OF OXYGENATED FUELS

The interagency report attempts to collect and interpret the existing data on the effects of winter oxygenate addition on emissions of pollutants other than CO. The attention devoted to this aspect is found in two short sections (1.32.2 and 1.59) and Appendix 3 of the interagency report. The main issue is whether addition of oxygenates to gasoline produces unintentional increases in pollutants other than the intended target CO reduction.

Measurements are quoted primarily from the Auto/Oil AQIRP study and from a few measurements of ambient data from selected locations during the period when oxygenated fuels were in use. The authors in general refer to the AQIRP bulletins (1991) rather than to the corresponding peer-reviewed SAE publications (SAE SP-920 and SP-1000 and additional publications) and sometimes refer to specific papers containing AQIRP data. It would be preferable to make specific reference to the peer-reviewed SAE publications.

Review of the existing data reveals a lack of measurements at conditions relevant for winter conditions (temperatures less than 35 °F) likely to be prevalent when the oxygenated-fuel program is in effect.

DYNAMOMETER TESTS

NO_x EMISSIONS

NO_x is a pollutant of major concern for areas not in attainment of the ozone National Ambient Air Quality Standards (NAAQS) (NRC, 1991a). Exceedances of the ozone NAAQS do not occur for many areas during the winter when oxygenated fuels are used to reduce ambient levels of CO. However, for areas not in attainment of ozone NAAQS, potential increases in NO_x from motor vehicles using oxygenated fuels should be given greater consideration in the interagency report.

The interagency report adequately covers the available data on the effect of oxygenates at the standard FTP conditions (75 °F); data from Table 1.3 of the interagency report indicate that NO_x emissions increase slightly with the addition of oxygenates (0 to 1% per weight % oxygen). Data from the AQIRP program for high emitters (high emitters were defined by the HC emissions relative to the average) indicate a more pronounced effect (up to 12% increase per weight % oxygen). However, conflicting data were provided by an EPA study (Mayotte et al., 1994a and b) showing that the effect of oxygenates on high emitters led to no or a small decrease in NO_x emission levels.

The conflicting data emphasize that, even at 75°F, the effects of oxygenates on NO_x emissions is complex. Figures 2.1 and 2.2 indicate that if the air-fuel ratio is rich and additional oxygen is supplied (regardless of the source) NO_x emissions increase and CO emissions decrease. It follows that, if the open-loop, air-fuel-ratio schedule of motor vehicles is constructed using a non-oxygenated fuel, use of an oxygenated fuel (with all other fuel characteristics the same) will increase engine-out NO_x. However, adding oxygenates may require changes in fuel composition to achieve the overall desired characteristics; these changes may affect NO_x. In addition, the net amount of pollutants emitted from a motor vehicle's tail pipe is the result of engine emissions modified by the effects of the catalytic converter. Consequently, NO_x emissions to the atmosphere are affected by a variety of factors including type of emission control systems, maintenance practice, and catalyst age.

More glaring is the lack of data at low temperatures. The reported data by Most (1989) in Table 1.6 of the interagency report cover the effect on CO, but no measurements are reported on NO_x. Additional data by Lax (1994) are provided in a review of a different study, remarking that "The NO_x emissions were generally increased by the addition of the fuel oxygenate" (last paragraph of page 12). The discussion on NO_x is not clearly separated in the current report, appearing as incidental comments in the discussion of CO effects, and again in the discussion of other emissions.

HC EMISSIONS, MTBE, AND TOXICS

The interagency report correctly cites the result of dynamometer studies: enleanment by addition of oxygenates in general decreases hydrocarbon emissions (Figure 1.1). Only one study is reported at low temperatures (Table 1.7), although presumably the report by Most (1989) in Table 1.6 also contains relevant data. The main conclusions quoted from the review by Hood and Farina (1995) appear to indicate that the available data fail to show consistent reductions of total hydrocarbon emissions with added oxygenates at low temperatures and that only modest reductions are achieved for properly operating, closed-loop control vehicles.

The effect of oxygenates on toxic emissions is mentioned only briefly. Clearly, very few data are available on the emissions of MTBE and aldehydes at low temperatures, even though a general decrease in total hydrocarbons is expected with the addition of oxygenated fuels, as well as a decrease (by dilution) in the emissions of benzene and 1,3-butadiene. FTP studies at 75°F indicate a 26% increase in emissions of formaldehyde (Gorse et al., 1991), and 5% and 8% decreases in benzene and 1,3-butadiene emissions with the addition of 15% MTBE to the gasoline. Ethanol addition leads to an increase in acetaldehyde emissions. These results are adequately reported in the conclusions to the emissions studies (p. 19) but are not carefully discussed in the main text. The authors should add a table summarizing the existing data on the effects of toxic compounds and total MTBE emissions. In addition, a link with the potential health effects of the different toxics relative to the different cancer potencies (e.g., a reduction in benzene but an increase in formaldehyde) in the health-effects section should be made.

FIELD MEASUREMENTS

There is clearly a gap in the available field data on emissions other than CO associated with oxygenate additions. No remote-sensing data on hydrocarbons or other emissions are apparently available. The authors of the report quote measurements by Anderson et al. (1993, 1994) indicating the increase in aldehyde and MTBE concentrations in ambient-air measurements in the Denver area after the introduction of the oxygenated-fuels program. The latter are not peer-reviewed publications and were not available to the committee. Another part of the report (HEI, 1996) indicates that measurements of MTBE, aldehydes, and other compounds are available for other areas of the country where oxygenated fuels have been in use. It would be useful to attempt to cross-reference or consolidate the few available air-quality measurements in one single section on ambient air quality-effects. Whereas most of the interagency report appears to be concerned with MTBE proper as a potential health hazard, it might be advisable to consider the combined effects of introduction of MTBE with the potential increase in aldehyde emissions and production of byproducts in the environment. Furthermore, the committee recommends that an assessment of the toxic emissions associated with the introduction of oxygenated fuels during the winter (or associated dynamometer tests at low temperatures) be made.

ATMOSPHERIC CHEMISTRY

The interagency report contains an appendix on the "Explanation of [OH] Trace Gas Life Time Estimates." The importance of secondary effects related to emissions of copollutants associated with oxygenated fuels deserves a discussion incorporated in the main report. In particular, a clearer understanding is needed of the atmospheric chemistry, transport, and fate of MTBE and byproducts including transport into waterways and production of secondary photochemical products. The discussion in Appendix 3 of the interagency report is a very good introduction to the subject and covers the existing literature, but the relatively long lifetime of oxygenated compounds raises some concern regarding the fate of the compounds.

FUEL ECONOMY, ENGINE PERFORMANCE, AND PROGRAM COSTS

There have been complaints that the addition of oxygenates to gasoline reduces fuel economy (measured in miles per gallon) and leads to poor engine performance. The interagency report addresses both of these issues.

After new oxygenated-gasoline programs were implemented in the winter of 1992-1993 to reduce CO emissions, consumers in some areas of the country expressed concerns that these fuels had led to large reductions in fuel economy and deterioration of engine performance. Losses in fuel economy in excess of 20% were claimed by some consumers. In addition, a variety of engine-performance problems, including rough engine operation and difficulty in starting, were reported. Therefore, it is important to establish the extent to which the addition of oxygenates to gasoline brought about these problems.

FUEL ECONOMY

The interagency report reviews a number of credible sources of data on the effect of addition of different oxygenates (MTBE, ETBE, and ethanol) on fuel economy. In general the studies cited[1] compare the fuel economy obtained with conventional fuels and that obtained with fuels containing oxygenates.

The report concludes that the reduction in fuel economy expected based on theoretical considerations is 2%-3% and that real-world changes correspond to these theoretical predictions. While the committee agrees in general with the conclusion, it recommends stating explicitly that fuel-economy changes are reliably predicted

[1]In all, 13 different studies or groups of studies are cited.

by the change in energy content per gallon of fuel brought about by a given change in composition[2]. There is agreement based on data from a wide variety of sources that if a given level of an oxygenate reduces the energy content per gallon of a formulated gasoline by 1.6%, for example, the expected reduction in fuel economy is also 1.6%.

Engine Performance

Fuel-related sources of engine-performance problems for all types of gasoline include excessively high or low volatility, water absorption, improper storage and handling, enleanment (higher oxygen-to-fuel ratios), reduced motor octane, and materials compatibility. These sources can lead to rough engine operation, overheating, damaged pistons, vapor lock, starting difficulties, plugged fuel filters, fouled spark plugs, fuel leaks, hesitation during acceleration, flooding, stalling, and engine fires.

The interagency report deals with all these issues and concludes that, except for possible drivability problems due to enleanment, performance problems due solely to the presence of oxygenates in gasoline are not expected. The committee agrees with this conclusion. As the report states, there are a number of factors other than gasoline quality and composition that contribute to engine-performance problems that may be corrected by relatively simple consumer actions.

Program Costs

The draft executive summary in the interagency report states that

---

[2]Energy content per gallon is the same as the heating value per gallon.

the report identifies "areas where the data are too limited to make definitive conclusions about the costs, benefits, and risks of using oxygenated gasoline." The committee believes that reasonably reliable costs of oxygenated-fuel production, fuel use, and program administration can and should be calculated to help guide the program. After calculating those costs, the interagency report might compare, at least at a broad level, the expected amount of ambient CO reduction using oxygenated fuels with the reduction that would be expected from alternative approaches, such as the repair or elimination of high-emitting vehicles for the same expenditure of funds.

MODERN TECHNOLOGY

In addition to increasingly precise adaptive control, a number of changes taking place are likely to change the outlook for CO emissions from gasoline-powered vehicles. The committee believes that the interagency group should consider these changes in its analysis.

Perhaps the most important factor is the change to 100,000-mile certification. Under this provision, auto manufacturers will be required to certify that vehicles will maintain emissions standards for 100,000 miles instead of the current 50,000 miles. In addition, vehicles will include onboard diagnostics that will detect malfunctioning emissions systems, so that consumers will know that repairs or adjustments are needed. These changes combined will result in better continued performance of these vehicles as they age in the population and should contribute to reductions in ambient CO concentration.

The sensors being used today in new cars have faster response during warm up. This reduces the time with open-loop operation; as a result, the time required to heat the catalyst becomes increasingly important in achieving a high level of CO control.

In addition, EPA is considering changes to the Federal Test Procedure used to specify the emissions from new vehicles that would take into account performance at low ambient temperatures. The current procedure measures emissions at an ambient temperature of 75°F, and these measures do not give a direct measure of CO emissions at lower starting temperatures, representative of winter operation in many areas. Modifying the FTP to include performance at winter ambient temperature would ensure that new vehicles would control CO emissions to specified levels under winter conditions, and this too would lower ambient CO concentration. Currently, motor vehicle manufacturers are not required to meet CO emission standards at temperatures less than 20°F.

The apparent advantage of a strategy to use oxygenated fuels is expected to dissipate with time as vehicles with advanced emission-control systems replace older vehicles. Closed-loop, adaptive-learning, and oxygen-sensor systems in advanced-technology vehicles control the fuel-combustion process at the same stoichiometry irrespective of whether the O_2 in the engine's cylinders comes from fuel or air. In such a case, oxygenation of fuel is expected to result in small changes in CO emissions.

OVERALL CONCLUSIONS ON AIR QUALITY, FUEL ECONOMY, AND ENGINE PERFORMANCE

The interagency report concludes that there have been substantial reductions in ambient CO levels in the last 20 years and that vehicle emission controls have been a major factor in this reduction. However, the federal report should better characterize the uncertainty about the extent to which oxygenated fuels have contributed to this reduction. The committee believes that it has not been established that oxygenated fuels have been a major factor in this reduction.

The committee agrees with the conclusion in the interagency report that, under many fuel-control systems, oxygenated fuels decrease CO emissions under FTP (at 75°F) conditions. However, the data presented do not establish the existence of this benefit under winter driving conditions.

The interagency report does not clearly address the effects of fleet composition, particularly high-emitting vehicles, on total CO emissions.

The interagency report highlights some differences between EPA's MOBILE 5a model results and observed vehicle emissions and ambient concentrations, but it does not provide an assessment of why those differences exist. It should do so and should also emphasize the fact that the model apparently overpredicts the oxygenated-fuel effect by at least a factor of 2 according to comparisons of model predictions of CO-emission reductions with observed data.

The interagency report provides a good summary of previous studies that have assessed the impacts of oxygenated fuels on winter CO concentrations; however, the report should state clearly that winter ambient CO reductions have been as low as zero and as high as about 10%.

The enleanment effect of oxygenated fuels presents the potential for increased NO_x emissions from motor vehicles. Furthermore, much of the available data suggests that such an increase does occur. Any increase in NO_x emissions could be detrimental in ozone non-attainment areas where exceedances have occurred during the period of the oxygenated fuels program.

The interagency report concludes that the fuel-economy penalty associated with the use of oxygenated fuels is approximately 2% to 3% and is related to changes in energy content per gallon. The committee agrees with these conclusions. The committee also agrees with the report's conclusion that engine performance is typically not adversely affected by their use.

OVERALL RECOMMENDATIONS ON AIR QUALITY, FUEL ECONOMY, AND ENGINE PERFORMANCE

The implications of the lack of emission data collected at low temperatures in evaluating the effectiveness of the oxygenated-fuels program in reducing CO emissions should be fully discussed in the interagency report.

A defensible field study with adequate statistical power should be performed at low temperature to assess whether oxygenated fuels reduce ambient CO concentrations under such conditions. If a beneficial effect of reduction in CO is demonstrated in that field study, a carefully designed study should then investigate the effects of oxygenated fuels on emissions of NO_x, VOCs, and toxic air pollutants on winter air quality, using appropriate controls and accounting for differences in such factors as fleet population, high-emitting vehicles, I/M programs, and local fleet characteristics. The introduction of toxic organic compounds into the air, as well as a fuel-economy penalty, should also be addressed in such a study. Also, it is equally important desirable to perform such a study under cold temperature conditions in an area where no oxygenated fuels are used. Results for an area using oxygenated fuels can be compared with areas not using oxygenated fuels.

Despite uncertainties in estimating costs and even greater uncertainties in estimating benefits, full evaluation of the oxygenated-fuel program requires that the interagency report address and document program costs and benefits at least at a broad level.

3

Water Quality

This chapter reviews the interagency report's discussion of fuel oxygenates and water quality. The purpose of that discussion was to address water-quality issues arising from the use of fuel oxygenates (primarily methyl tertiary-butyl ether (MTBE)) and their movement in the hydrologic cycle. In general, the interagency report provides a clear and factual presentation of our current understanding of MTBE fate and transport in the environment. However, the water-quality section of that report does not adequately emphasize that there is only a small amount of data available on MTBE—a very widely used, potentially hazardous industrial chemical. The large majority of states do not have any current programs or requirements to monitor concentrations of MTBE or other fuel oxygenates in water. Yet the available data suggest that MTBE is sometimes present in precipitation, stormwater runoff, groundwater, and drinking water. The interagency report should clearly recommend that state and federal agencies immediately begin monitoring for MTBE and related oxygenates immediately. Until these data are collected, we will not be able to evaluate nationwide the range of aquatic-resource contamination or the exposure of humans to fuel oxygenates.

The limited data available suggest that currently the waterborne exposure pathway of greatest concern is consumption of groundwater contaminated by point-source releases from leaking underground storage tanks (USTs). It is common practice in those states that currently monitor MTBE to close or treat water supplies contaminated with higher concentrations of this compound, and thus exposures via this pathway are expected to be minimized. In the majority of states that do not monitor for MTBE, it is not known to what degree individuals are exposed to higher levels of MTBE via drinking water.

Monitoring data show that releases of MTBE through nonpoint sources (e.g., precipitation and small surface spills) can enter storm water-runoff and shallow groundwater and potentially enter drinking-water supplies. At present, the relative risk of exposure to point and nonpoint sources of MTBE are not adequately characterized. MTBE concentrations in water associated with nonpoint-source releases need to be carefully monitored to determine whether these concentrations are increasing with continued use of MTBE. In addition, the effect of landuse, storm-water-management practices, and hydrogeologic factors on MTBE concentrations should be assessed.

OVERVIEW OF WATER QUALITY IN THE INTERAGENCY REPORT

The interagency report provides a summary of major sources of fuel oxygenates in the hydrologic cycle; monitoring data on MTBE concentrations in surface water, storm-water runoff, and groundwater, including drinking-water supplies; and a review of the major processes expected to influence MTBE transport and fate in the environment. Most of the MTBE monitoring data presented in the interagency report were obtained directly from data collected by regulatory agencies or by the U.S. Geological Service.

Oxygenates added to gasoline have the potential to degrade water quality due to environmental releases during gasoline production, handling, and use. MTBE is the most commonly used oxygenate, followed by ethanol (EtOH). Much lower amounts of ethyl tertiary-butyl ether (ETBE), tertiary-amyl methyl ether (TAME), diisopropyl ether (DIPE), and methanol (MeOH) are also used. Monitoring of groundwater, storm water, and drinking water have shown the presence of MTBE and have raised questions about the extent to which MTBE and possibly other oxygenates might move through the hydrologic cycle.

Historically, the primary concern over oxygenate contamination in water has been associated with substantial surface spills and leakage from USTs. Because of the oxygenates' high solubility in water, they can partition readily from gasoline into water, resulting in high aqueous concentrations. In shallow groundwater down-gradient of USTs, MTBE concentrations up to 200,000 μg/L have been observed (Davidson, 1995). EtOH is more soluble in water than MTBE and may occur in higher concentrations in groundwater and surface water. However, compared to MTBE, EtOH is biodegraded more rapidly under a wide range of environmental conditions. MTBE is much more difficult to biodegrade and likely to persist when introduced into groundwater or surface water.

Recent results from monitoring of storm water and shallow groundwater have raised concerns about the potential for diffuse or nonpoint sources of MTBE. In studies of 16 U.S. cities with populations greater than 100,000, the interagency report indicates that MTBE was detected in storm-water runoff in eight cities. Of these eight cities, only three had an oxygenated-fuels program in place at the time of sampling. In the other five cities where MTBE was detected, it was likely used as an octane enhancer in gasoline.

When data from all 16 cities were examined, MTBE was detected in 6.9% of the 592 samples analyzed. When detected, the concentration ranged from 0.2 to 8.7 μg/L. (The limit of detection of MTBE in water reported in these studies was 0.2 μg/L.)

MTBE present in storm-water runoff is hypothesized to originate from two sources: (1) contact of surface runoff with minor surface spills associated with petroleum refueling; and (2) partitioning of MTBE out of the ambient air into precipitation. The half-life of MTBE in the atmosphere is longer during winter months than in warmer months, and MTBE is expected to build up to somewhat higher concentrations in the atmosphere. At lower temperatures, there is an increased tendency of MTBE to partition from the air into water (Squillace et al., 1995a). Both these factors likely combine to cause higher concentrations of MTBE in winter precipitation and winter storm-water runoff. While no direct measurements of MTBE in precipitation are available, according to the interagency report the frequency of MTBE detection in the 16-city stormwater-monitoring study was substantially higher in the winter than in the summer.

In cases where precipitation or storm-water runoff contains MTBE or other oxygenates, there is a potential for contamination of shallow groundwater supplies to result in low concentrations of these compounds. MTBE is not routinely monitored as part of assessments of groundwater when no contamination is expected. However, in the few studies that did look for MTBE, the interagency report indicates that MTBE was detected at low levels (0.2 - 10 μg/L) in 13% of the 540 wells sampled. Because of the relatively low MTBE levels detected and the physical location of the wells, the report indicates that these detections were not due to releases from concentrated spills or UST leakage. The measurement of low concentrations of MTBE at scattered locations was used as an indication of diffuse contamination of shallow groundwater by contaminated precipitation or storm-water runoff.

At present, the potential health and environmental impacts of oxygenate contamination in storm-water runoff and shallow groundwater are not adequately characterized. EPA had published a draft lifetime health advisory with a lower limit concentration, 20 μg/L, of MTBE in drinking water. (The interagency report mentions that the health advisory for MTBE is expected to be revised in 1996.) For the small number of water supplies sampled, high levels (exceeding the 20-μg/L health advisory) were rarely observed. However, the interagency report indicates that MTBE concentrations between 0.2 and 10 μg/L were occasionally observed in drinking-water supplies. If EPA's recommended health-advisory concentrations were reduced considerably, substantial concerns would arise about the potential hazards of MTBE in drinking water.

The impacts of storm-water and groundwater containing oxygenates on aquatic life are very poorly understood. Although acute-toxicity assays have been performed for several important species, chronic-toxicity data are lacking. Without chronic-toxicity data, it is not possible to determine whether important aquatic biota might be impaired by exposure to MTBE or other oxygenates.

The interagency report indicates that the concentration of MTBE in surface water will normally be reduced by volatilization, which occurs more rapidly when the water is shallow and the overlying atmosphere contains little or no MTBE. In the subsurface, transport of MTBE will be influenced by exchange with the gas phase and possibly by biodegradation. Sorption to sediment or aquifer material is not believed to be an important attenuation mechanism, because of the high solubility and low octanol-water partition coefficient of MTBE and other oxygenates.

At this time, the influence of biodegradation on the transport of MTBE through the subsurface is not well understood. While a few studies have shown limited MTBE biodegradation, it has occurred only under restricted environmental conditions (Daniel, 1995;

Thomas et al., 1988; Mormile et al., 1994; Yeh and Novak, 1995). Also, in most biodegradation studies, MTBE biodegraded in only a small fraction of the total samples examined, suggesting that MTBE biodegradation may not be common in the environment. The presence of MTBE or other oxygenates does not appear to affect the biodegradation of other fuel contaminants adversely in most cases (Hubbard et al., 1994). However, because of MTBE's recalcitrance to biodegradation, the presence of MTBE in the fuel mixture might reduce the applicability of intrinsic bioremediation or natural attenuation for the management of UST releases. The importance of abiotic degradation processes or other nonbiological processes (e.g., chemisorption) are not addressed at all in the interagency report.

SECTIONS OF THE INTERAGENCY REPORT REQUIRING CLARIFICATION OR REVISION

The interagency report should be revised to state clearly the actual sources of the monitoring data used in the analyses. It is not clear in the report whether the data presented were obtained directly from the raw data files or from summary reports. For example, it is not clear whether the groundwater-monitoring data were obtained directly from the National Water Quality Assessment (NAWQA) files or from summaries prepared by Squillace et al. (1995).

Nearly half the reported detections of MTBE in groundwater were from the Denver-area NAWQA study. These data might not be representative of nationwide trends, for two reasons. First, a number of the wells sampled in this study were up-gradient wells monitored during UST investigations where the MTBE detection could have been related to a UST release, not nonpoint-source contamination. Second, dry wells and other infiltration devices are often used in the Denver area for storm-water management. These devices should rapidly introduce contaminants into the subsurface and may greatly increase the likelihood of groundwater contamination by storm-water runoff. While Denver is not the only area of the country that uses these devices, in a great many areas lower-permeability soils restrict the use of infiltration measures for storm-water management.

The interagency report should discuss the current understanding of the extent to which abiotic degradation mechanisms or other nonbiological mechanisms (e.g., chemisorption) reduce groundwater concentrations of MTBE and other alkyl ether oxygenates.

The rate of degradation of MTBE and other oxygenates in the atmosphere can have a substantial influence on ambient concentrations in air and precipitation. On page 28 of the water-quality chapter of the interagency report, the half-life of MTBE is reported to be 4 days at 25°C. However, in the appendix of the air-quality chapter, the winter-time half-life of MTBE is shown be vary from 20 to 300,000 days depending on latitude. This apparent discrepancy needs to be resolved.

The rate of volatilization from surface water will be controlled by the concentrations in the aqueous and gas phases and the mass-transfer efficiency. The interagency report should note that when the overlying atmosphere contains significant concentrations of alkyl ether oxygenates (AEOs) and aqueous concentrations are low (such as might occur during overland flow), the extent of volatilization might be negligible.

The discussion of volatilization effects on AEO transport to groundwater needs to be revised substantially with emphasis on the uncertainties in this area, as well as on documentation of the modeling approach, and assumptions, boundary and initial conditions, and major model parameters. The text should also clearly distinguish between aqueous and gas-phase diffusion processes. This section of the report assumes a relatively constant level of AEOs in the atmosphere, which would result in a gradual transfer of contaminants to groundwater. A plausible alternative scenario is one in which atmospheric AEO concentrations increase in the winter and then decrease in the summer. Under this scenario, soil-water AEO concentrations might increase in the wintertime due to high atmospheric concentrations and then decrease during the summer due to diffusive losses back to the surface, resulting in a very low net flux of AEOs to the water table.

DISCUSSION OF INTERAGENCY REPORT RECOMMENDATIONS

The interagency report recommends that an ad hoc panel representing public and private sectors be formed to develop a comprehensive research and assessment plan to determine the impact of alkyl ether fuel oxygenates on drinking-water quality and aquatic life. Development of this plan is certainly worthwhile, but should not be a reason to delay implementation of the other recommendations in the interagency report. The interagency report has already made a substantial contribution toward defining what is known about sources and transport of alkyl ethers in the hydrologic cycle. Government agencies and industry should immediately begin work to fill the major information gaps identified in the interagency report.

The interagency report recommends that the AEOs e.g., MTBE, ETBE, TAME, and DIPE—be added to existing VOC analytical schedules and as routine target analytes for VOCs in drinking-water, wastewater, surface-water, groundwater, and remediation studies.

The committee supports this recommendation and suggests that these compounds also be added to the list of target analytes in EPA methods for VOCs in air. Addition of AEOs to the air and water analyte lists could be accomplished at little additional expense and would greatly improve our understanding of the sources and fate of these compounds in the environment.

The interagency report recommends that a national database be developed to catalog analytical determinations of AEOs in air and water. This database could then be used in exposure assessments for AEOs in drinking water and for aquatic life in surface water.

Over the short term, it is unlikely that sufficient additional information will be developed that would allow accurate exposure assessments of AEOs. Over the longer term, these data can be more efficiently collected and compiled through existing monitoring programs. Consequently, there appears to be little benefit in developing a separate database on concentrations of AEOs in different media at this time. However, it would be very useful to identify a lead agency or organization that would continue to compile information as it becomes available on the occurrence, transport, and fate of AEOs in different media. As analytical methods for measuring these compounds come into more widespread use, the pool of monitoring data for AEO concentrations should rapidly increase. At that time, it may be useful to revisit this issue and develop a computerized database for use in risk assessment.

The interagency report recommends that a database be developed on AEO use in different cities and regions to identify correlations between seasonal use and water quality.

At this time, it is not clear what would be the benefit of demonstrating that AEO concentrations did or did not correlate with use. In the absence of a clear benefit, development of the administrative infrastructure required to collect and organize these data on a national scale does not appear to be warranted.

The interagency report recommends that annual releases of AEOs to the environment from all sources (e.g., industrial releases, refueling losses, auto emissions, and storage-tank releases) be determined to identify the primary sources of AEOs in the environment and aid in setting priorities for further reduction.

With currently available information, it will probably not be possible to develop precise estimates of AEO releases. Over the short term, efforts should be focused on developing order-of-magnitude estimates of releases from likely AEO sources. These estimates will help to rank sources for potential reduction and to identify major data gaps that will need to be eliminated before more precise emission estimates can be developed. Over the long term, development of more accurate estimates of releases will be needed if we are to improve our understanding of the fate and transport of AEOs throughout the hydrologic cycle.

The interagency report recommends that the available data be reviewed annually to assess the effects of AEO-use, land-use, and hydrogeologic variables on temporal and spatial variations in water quality .

While a continual assessment of the effects of oxygenate use on water quality is very important, it is not clear whether this assessment should be conducted as a separate program or should be integrated into existing water-quality assessment and management systems. Each state is now required to perform a biannual assessment of surface-water, groundwater, and drinking-water quality and the major factors influencing water quality (305b Reports). This information is incorporated into a national water-quality assessment report prepared by EPA.

Once AEO measurements are incorporated into the major analytical protocols, the states will begin to accumulate a wealth of information on sources and transport of these compounds in various compartments of the hydrologic cycle. In the short term, these data should be compiled to update the interagency report. Over the longer term, the states should be encouraged to address the effect of AEOs on water quality and associated beneficial uses in the 305b reports. The federal government's primary role might best be to integrate this information and to support additional targeted studies that address regional and multimedia issues.

The interagency report recommends that large-scale monitoring studies be performed in selected areas to characterize the concentrations and mass fluxes of AEOs between the different compartments of the hydrologic cycle (atmosphere, surface water, and groundwater); that laboratory and field studies be conducted to identify factors controlling the degradation of AEOs in the environment, the rate of degradation, and potential degradation products; and that theoretical modeling be performed to evaluate factors influencing the transport of AEOs from the atmosphere to shallow groundwater.

The committee supports the need for additional research in all these areas. A portion of this work would probably be done most effectively by individual investigators. However, for this topic, coordinated multi-investigator projects could yield substantial benefits. Ideally, several cities or regions with differing climatic, land-use, storm-water management, and hydrogeologic conditions would be selected for multimedia studies of AEO transport and fate in the environment. Through proper coordination with industry, it should be possible to estimate seasonal use of AEOs and subsequent air emissions in these target areas. These emissions may then be compared with AEO concentrations in ambient air, precipitation, storm-water runoff, surface water, and groundwater. By conducting coordinated large-scale monitoring projects, small-scale field studies, and controlled laboratory experiments, it should be possible to evaluate current theories and mathematical models of AEO fate and transport.

The interagency report recommends that coordinated chronic- and acute-toxicity studies of AEOs be performed with a broad range of aquatic animals and plants to assess the threat to aquatic life and form a basis for federal water-quality criteria.

The committee supports this recommendation. Intrinsic bioremediation is rapidly becoming accepted as a cost-effective, environmentally acceptable strategy for managing releases from USTs.

Through this approach, naturally occurring microorganisms are allowed to degrade petroleum constituents, such as benzene, toluene, ethylbenzene, and xylenes. However, MTBE and other AEOs are believed to be more resistant to biodegradation and are assumed to be resistant to abiotic degradation processes. Thus those compounds might be transported to nearby streams. States and other regulatory bodies now need aquatic-toxicity data to determine whether discharge of AEOs to surface waters will adversely affect aquatic resources.

COMMITTEE'S CONCLUSIONS

The large majority of states do not have any requirements in place to monitor MTBE or other fuel oxygenates in storm-water runoff, groundwater, or drinking water. The absence of these monitoring data prevents an accurate assessment of exposure of humans or aquatic biota to MTBE and implementation of control measures to prevent adverse impacts.

On the basis of the small amount of monitoring data, MTBE has been detected in less than 5% of the groundwater samples analyzed, suggesting that drinking water is not currently a major MTBE exposure pathway for much of the population.

RESEARCH NEEDS

Storm-water runoff and shallow groundwater can be contaminated with low levels of MTBE (<20 μg/L) via precipitation or contact with small surface spills. These contamination sources should be carefully monitored to evaluate changes over time and the effect of land use, storm-water management practices, and hydrogeologic factors on MTBE concentrations in environmental media. If EPA considerably lowers the level of its recommended health-advisory concentration for MTBE, substantial concerns would arise about the potential for nonpoint sources of MTBE to affect water supplies adversely.

More needs to be known about the biodegradation of MTBE and other AEOs in surface water, soil, and groundwater. Biodegradation processes have the potential to substantially reduce the effects of point- and nonpoint-source releases of MTBE and other oxygenates. Current information should be assessed to determine whether a better understanding of abiotic degradation is an important research need.

4

Human Exposure

To understand the risk associated with use of oxygenated fuels, it is necessary to understand the exposure pathways and characteristics of human contact with oxygenated fuels and the products of their combustion in motor vehicles. Multiple pathways exist for exposure to gasoline and its components, including occupational contact with fuel components through product distribution and use. The general public is exposed during refueling and use of gasoline products. Exposures can also occur from the environmental transport and transformation of gasoline constituents released as evaporative or tailpipe emissions from in-use vehicles.

The major pathway of exposure to oxygenated fuels and other gasoline products is assumed to be inhalation. However, ingestion and uptake through dermal contact might also be important in some cases. For example, occupational exposures of service-station attendants, automobile mechanics, and distribution workers may involve dermal as well as inhalation exposures. Similarly, the general public has the potential for dermal exposure during the handling and use of gasoline products, such as while cleaning and degreasing. Ingestion may also be an important pathway of exposure for individuals using well or surface waters contaminated by oxygenated fuels. Well- and surface-water contamination has not been effectively characterized to date.

The concentration, duration, and frequency of contact with oxygenated fuels are vital components of exposure; they influence uptake of a substance by the body, resulting doses to target organs, and health effects. Potentials for acute effects and symptoms can be influenced by the duration and frequency of contact with the contaminant. In general, higher concentrations encountered for shorter periods (e.g., during refueling) tend to be responsible for acute effects, and longer-term low-level concentrations are generally associated with chronic health effects.

The use of oxygenated fuels is designed to reduce exposures to CO emitted from motor-vehicle tailpipes. Such fuels might increase or decrease the air concentrations of organic toxics associated with evaporative or tailpipe emissions (e.g., benzene and formaldehyde). Therefore, comprehensive risk evaluation would require a comparison of the risks resulting from shifting exposures from environmental contaminants of conventional gasoline to those of oxygenated fuels.

DATA REVIEWED BY THE INTERAGENCY REPORT

The interagency report contains two documents referred to as HEI and OSTP reports. Each of these reports has sections on exposure assessment but takes a different approach to the issue. The primary focus of both reports is on inhalation exposures; substantially less attention has been devoted to the potential for dermal exposures and direct ingestion of oxygenated-fuel components. As stated above, these routes of exposure might also be important in some cases.

The HEI report reviews occupational- and nonoccupational-exposure studies and presents a summary of the range of concentrations and exposures observed in the literature. To date, there have been no studies that are representative of a defined sample of the population. The reported studies indicate concentrations that span 5 orders of magnitude and generally decrease with increasing averaging periods. These results are presented in the HEI report, which concludes that the existing data provide a rough estimate of exposure ranges associated with various activities for the general population, [while] the frequency and distribution of these activities and the amount of exposure by dermal and oral routes are uncertain. Because of these limitations, using these data to calculate a cumulative exposure for use in risk assessment is not appropriate.

The OSTP report does not heed this warning and uses the same set of exposure studies evaluated by HEI to estimate cumulative exposures for two hypothetical exposure scenarios that are created to evaluate lifetime cancer risk. The exposure section of the OSTP report makes it clear that the cumulative exposure estimates are based on hypothetical scenarios, with one of the two designed to represent a "reasonable worst-case" scenario. However, the limitations can easily be missed by a reader in the subsequent risk characterization, in which the worst-case scenario is coupled with potency estimates to calculate lifetime cancer risks. The exposure characterizations in the HEI and OSTP reports are critiqued below.

COMMITTEE CRITIQUE

OSTP REPORT

The OSTP report reviews several studies of MTBE exposure, including ambient studies, an in-vehicle study, and several occupational studies. These studies present results from samples of personal or microenvironmental exposures that were not selected as representative samples from the cities in which monitoring was conducted. They provide a general view of individual exposures and provide some information on the range of concentrations that might be observed in areas where MTBE is used. Point estimates of concentrations from the microenvironmental measurements are combined with estimates in the OSTP report to provide cumulative estimates of exposure.

To estimate potential human exposures, two exposure scenarios of hypothetical exposure sequences are constructed. Scenario I is based on a person who visits a gasoline station 1.5 times/week, commutes 10 h/week, visits an auto-repair shop 4 times/yr, and spends 57 h/week in an office or public building. The home of the person in Scenario I is assumed to have a detached garage and not be near a gasoline station or highway. Scenario II is similar, with the exception that (1) the second hypothetical person is exposed to gasoline refueling concentrations that are 10 times higher, (2) a garage is attached to the home of this individual, and (3) the person is assumed to spend time outdoors near a gasoline station or heavily traveled highway. With these microenvironmental assumptions, Scenario II is assumed in the OSTP report to represent a reasonable worst-case exposure.

Both scenarios are arbitrarily defined but may help to bound potential high-end lifetime exposures. Microenvironmental concentrations used in these scenarios are rounded up to the next highest half order of magnitude in order to provide a high estimate of exposure. However, the committee noted that lifetime exposures derived for both exposure scenarios exceed the range of 24-h exposures estimated on the basis of empirical data presented in the HEI report. The OSTP-report scenarios, therefore, are exceedingly conservative. While it is often useful in a data-limited situation like this to generate hypothetical scenarios as distributional exposure data become available, these data should be incorporated into the exposure distributions.

The OSTP report states that there are limitations to estimating exposure, because of limitations in the models relating automobile emissions to ambient-air quality; and, for purposes of risk characterization, Scenario II is assumed for a lifetime.

HEI REPORT

The HEI report presents a useful summary of the studies reporting on occupational and nonoccupational measurements of personal and microenvironmental exposures to MTBE. These studies report on convenience (nonprobability) samples of exposure and were collected for different averaging times. Therefore, they are not suitable for *direct* use for construction of probabilistic exposure distributions in a risk assessment. However, the collective data, as presented in Figure 4-1, do provide information on the range of concentrations observed for different time averages.

These exposure data, however, can be used to provide a "reality" check for quantitative risk assessments and to construct boundary conditions or exposure ranges for MTBE concentrations that might be encountered in a variety of locations and activities. Using this information, the committee concluded that HEI could have performed a quantitative risk assessment using the exposure data represented by the *median* (approximately 0.13 ppb) or *maximum* (approximately 0.01 ppm) daily MTBE exposures as estimates of exposure. Both of the hypothetical lifetime-exposure scenarios derived in the OSTP report lie above the range of 24-h concentrations presented in the HEI report. The lower of the two OSTP lifetime-exposure scenarios used an average daily MTBE exposure of 0.018 ppm during the oxygenated-fuel season. This hypothetical scenario was generated by accumulating time-weighted concentrations assumed to exist in multiple microenvironments (e.g., refueling, commuting, at work, and at home). Scenario II, constructed as a hypothetical worst case was even further outside the range of

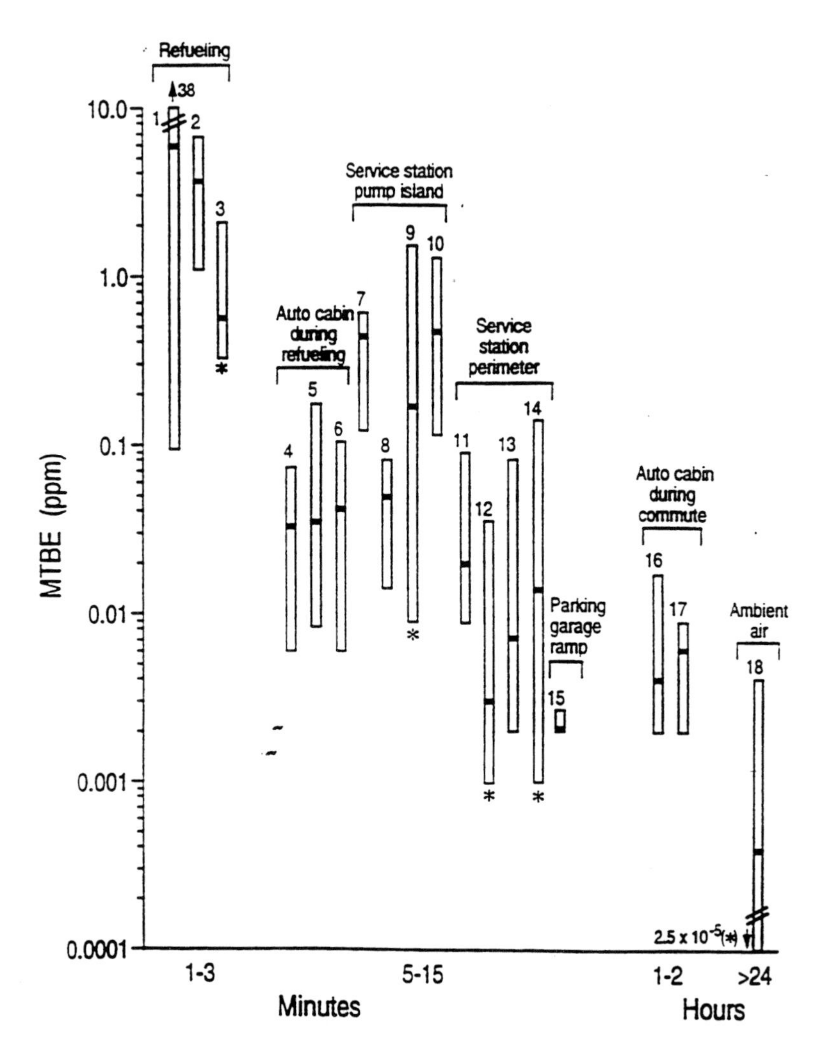

Figure 4.1. Time-weighted average exposures of individuals in the general public to MTBE. Only studies that provided ranges of exposure levels are included. The solid lines across the bars indicate median values. The numbers at the tops of the bars correspond to the numbers in parentheses in the "Sampling Site" column of Table III.3 (of HEI (1996)), where the same data are expressed. An asterisk (*) denotes the minimal detectable concentration. Source: HEI, 1996.

observed 24-h concentrations presented in the HEI report. In a data-limited situation like this, bounding scenarios are appropriate for conducting risk analyses. The scenarios, however, should be realistic, and the limitations to generalizability need to be stressed wherever the scenarios are used.

In addition to MTBE exposures, the HEI report provides a qualitative discussion of potential changes in concentrations of the atmospheric transformation products of MTBE, CO, air toxics, and O_3 precursors. The general direction of change for these gases is discussed, but the report argues that data are not sufficient to present a quantitative evaluation of the shift in concentrations. The report concludes that the impact of oxygenates on ambient CO concentrations has not been fully evaluated.

A comparative risk assessment requires data on the influence of MTBE on human exposure to these other constituents, and this committee believes that a quantitative framework should and could be established for this evaluation. Establishing a framework for consideration of all exposures will help identify important data gaps and provide for a direct comparison of the risk and benefits between conventional and oxygenated fuels.

CONCLUSIONS

- The committee finds important deficiencies in the OSTP exposure analysis, which calculates cumulative-exposure estimates for two hypothetical scenarios. The lifetime exposures calculated for these scenarios are 10 times higher than the maximum daily exposures based on empirical data that are summarized in the HEI report. In addition to a reasonable worst-case scenario, OSTP should generate a more-realistic baseline exposure. Other emission products should also be considered.
- The HEI report provides a comprehensive and useful summary

of MTBE-exposure studies. Concentration ranges encountered in occupational and nonoccupational situations are adequately represented. HEI expressed the opinion that the data are too limited to calculate cumulative exposures for risk assessment. However, the committee's opinion is that the data are sufficient to bound a quantitative risk analysis and to develop a framework for conducting a comparative risk assessment of conventional and oxygenated fuels. A quantitative framework should also be established for evaluating changes in exposure from related emission products.

RESEARCH NEEDS

- Routine ambient monitoring of MTBE and one of its major products of photo-oxidation, tertiary-butyl formate, should be initiated in communities where MTBE is used. (TBF has the potential to accumulate and persist in the atmosphere and its toxicity is unknown.)
- Representative personal-exposure monitoring of MTBE in an exposed population is needed in order to describe the distribution of exposures and for input into risk analyses. Such exposure monitoring should include the characterization of each individual's time-activity patterns, especially in the microenvironments where important exposures are likely to occur (NRC, 1991b). The most important factors affecting personal exposure should be determined in such a study.
- The relationship between fixed-site community monitoring and personal exposures to MTBE and other pollutants of concern resulting from the use of oxygenated fuels, should be evaluated in order to determine the information value of using community monitoring for assessing human exposure.

5

Potential Health Effects of Oxygenates

In evaluating the potential adverse health effects of oxygenated fuels, the committee was charged with reviewing three reports and providing critique regarding scientific credibility, comprehensiveness, and internal consistency of the data presented. In addition, as part of its charge, the committee was to identify priorities for research to fill data gaps. The reports to be reviewed included the Health Effects Institute's Oxygenates Evaluation Committee's report *The Potential Health Effects of Oxygenates Added to Gasoline: A Review of the Current Literature* (February 1996), the HEI report; the National Science and Technology Council Committee on Environment and Natural Resources and Interagency Oxygenated Fuels Assessment Steering Committee Report *Interagency Assessment of Potential Health Risks Associated with Oxygenated Gasoline* (February 1996), the OSTP report; and a memorandum from the Centers for Disease Control and Prevention signed by Richard J. Jackson and directed to the Interagency Oxygenated Fuels Assessment Steering Committee dated March 12, 1996, the CDC white paper, which compared and contrasted the other two reports.

Both the HEI report and the OSTP report relied upon published, peer-reviewed literature; unpublished reports from a number of sources, including industry, government agencies, and scientists; and personal communications. The National Research Council's Committee on Toxicological and Performance Aspects of Oxygenated Motor Vehicle Fuels was not provided with the original sources of data from which the review reports were written, so a full and complete critique of scientific credibility, comprehensiveness, and internal consistency of the data is not possible.

METABOLISM, DISPOSITION, AND TOXICOKINETICS

DATA REVIEWED IN THE HEI AND OSTP REPORTS

The HEI report reviews what is known about the metabolism, disposition, and toxicokinetics of MTBE and ethanol in both animals and humans. Some recent data on the disposition of ethyl-tertiary-butyl ether (ETBE) in animals are also reviewed. The report states that there is essentially no information available on the disposition, metabolism, or toxicokinetics of other proposed oxygenates. As is summarized in the report, following absorption, the major pathways for elimination of MTBE are exhalation and oxidative demethylation to form tert-butyl alcohol (TBA) and formaldehyde. MTBE metabolism displays saturation kinetics. TBA appears to have a longer half-life in the body than MTBE, and thus TBA in blood may be a better biologic marker of exposure than MTBE blood levels. Formaldehyde has a very short half-life in the body. Thus, its potential increase in blood after exposure to MTBE has not been studied. Differences between MTBE levels and the ratio of MTBE to TBA in blood following chamber and field studies were noted, and it was postulated that this could be due to differences in sampling times.

With respect to ethanol, HEI concludes that at the expected levels of exposure to ethanol in oxygenated fuels, the incremental blood levels of ethanol are predicted to be insignificant compared with endogenous blood levels. It is also unlikely that the fetus would be affected by this insignificant increment.

COMMITTEE CRITIQUE

There do not appear to be any major differences between the HEI and the OSTP report in regard to disposition, metabolism, and toxicokinetics of MTBE, and the committee is in basic agreement with the review and evaluation of the data presented, with the following exception.

Although the disposition of inhaled and orally administered radiolabeled MTBE in animals is reviewed in the HEI report, there is no definitive discussion of the amount absorbed into the systemic circulation following either route of exposure. However, it is stated that the data available from studies in which human volunteers were exposed to MTBE in chambers "suggest that less than half of the MTBE administered (32% to 42%) is absorbed after inhalation" and that this is consistent with the animal studies.

For purposes of extrapolating health-effects data from animals to humans, knowing the percentage absorbed following inhalation and ingestion for both animals and humans would be very useful. The final report needs to address whether this information is available from the studies already conducted, or if more studies are needed to obtain this information.

CONCLUSIONS

The committee is in basic agreement with the review and evaluation of the data presented in the HEI and OSTP reports. No major differences between the HEI and OSTP reports were noted with regard to the metabolism, disposition, and toxicokinetics of MTBE. The committee does note that the final report should address whether or not available animal studies address the percentage absorption following inhalation and oral exposures. This information would be useful when extrapolating health-effects data from animals to humans. The committee also concludes that it is essential that the actual exposure to MTBE be better quantified in any future epidemiological studies designed to evaluate the possible acute effects of inhaling MTBE. It is essential that the kinetics of MTBE and TBA in blood be established so that they can be used as biologic markers of exposure. In this regard, it would also be useful if measurements of formate (the metabolized product of formaldehyde exposure) from inhalation of MTBE were compared to endogenous formate levels.

RESEARCH NEEDS

The HEI report makes a number of research recommendations for additional studies on the metabolism and disposition of MTBE and other oxygenates and notes that some work is already under way. As noted above, the committee concludes that additional research on the metabolism and disposition of MTBE would be very useful as a basis for extrapolating health-effects data from animals to humans (i.e., a physiologically based pharmacokinetic (PBPK) model). Research should also focus on determining a suitable biologic marker of exposure for use in any future epidemiological studies on the acute effects of MTBE-gasoline mixtures. It would be prudent to conduct deposition and metabolism studies on other oxygenates before they are introduced and extensively used.

SHORT-TERM HEALTH EFFECTS

ANIMALS

DATA REVIEWED BY THE HEI AND OSTP REPORTS

The HEI report presents a thorough review of the animal studies that have been conducted on MTBE, ethanol, and other oxygenates. In animals, essentially all these compounds cause sedation and other reversible CNS effects, such as loss of motor coordination and decreases in motor activity. None of the compounds that have been studied appear to be irritants to the eye or upper respiratory tract or cause pulmonary irritation even at very high levels (i.e., 4,000-8,000 ppm for MTBE). Most ethers studied also appear to cause neurotoxic effects following exposures to high concentrations. For most effects, the no-observed-adverse-effects level (NOAEL) is at least 1,000 times higher than predicted human exposures.

One major difference between the HEI and OSTP reports is that the HEI report states that a NOAEL for motor activity was not achieved in inhalation studies in which rats were exposed to MTBE. The HEI report emphasizes the alteration in motor activity observed in rat inhalation studies was reversible and that no neuropathologic condition was observed. The CDC white paper discussing differences between the two reports states that OSTP examined the raw data from the study in question and concluded that the effect on motor activity at 800 ppm was not significant when examined with appropriate statistics which corrected for multiple comparisons. If this is the case, there would not appear to be a need for additional studies.

COMMITTEE CRITIQUE

One area not addressed in either the HEI or the OSTP report is the question of TBF toxicity. TBF is one of the major photo-oxidation byproducts of MTBE. If TBF levels are increased in the ambient air due to MTBE use, the available literature on TBF toxicity needs to be reviewed in the report. Based on that review and the extent of the exposures predicted, additional studies to characterize the acute and chronic toxicity of TBF may need to be added.

Based on the available data, the committee is skeptical about the need for additional studies based solely on the motor-activity studies conducted in rats. The committee notes that even at 800 ppm the effect, if any, is at an exposure of 100-1,000 times greater than predicted human exposures. It is also a reversible effect characteristic of this class of compounds and there is no indication of neuropathologic condition or persistent neurotoxicity following exposure to MTBE or other ethers.

CONCLUSIONS

The HEI report represents a thorough review of the short-term animal studies that have been conducted on MTBE, ethanol, and other oxygenates. However, one area not addressed in either the HEI or the OSTP report is the question of TBF toxicity.

Based on the available data, the committee does not feel the effect of MTBE on motor activity reported in rat inhalation studies is of major concern, for the following reasons: (1) the effect occurs at 100-1,000 times predicted human exposures, (2) the effect is reversible, and (3) there is no indication of neuropathologic condition or persistent neurotoxicity following exposure to MTBE or other ethers, even after long-term exposure to high levels of MTBE.

RESEARCH NEEDS

The HEI report notes the need for additional short-term animal studies on MTBE that determine blood levels in addition to evaluating CNS function. The committee agrees that such studies would be useful because measurements of blood levels of MTBE and its metabolites in animals, above and below the no-observed-effects levels for CNS effects, could be compared to blood levels expected in people (from PBPK modeling) following occupational or environmental exposures. In particular, adequate studies on other proposed oxygenates should be conducted before they are extensively introduced into the U.S. gasoline supply.

HUMAN

EPIDEMIOLOGIC STUDIES

ALASKA STUDIES

DATA REVIEWED IN THE HEI AND OSTP REPORTS. The HEI report and the OSTP report reviewed four studies: (1) the cross-sectional studies conducted in Fairbanks (Beller et al., 1992) and (2) Anchorage (Chandler and Middaugh, 1992); (3) the longitudinal study which was conducted in Fairbanks by the State of Alaska Department of Health and Social Services and CDC in the winter of 1992-1993 (CDC, 1993a; Moolenaar et al., 1994); and (4) a retrospective study that compares outpatient insurance claims made in Alaska during the winter months of 1992-1993 (i.e., months in which oxygenated fuels were in use) to claims made during the same months of previous winters (Gordian et al., 1995). The OSTP report reviews the same four studies as the HEI report.

COMMITTEE CRITIQUE. The OSTP report provides more detail

than does the HEI report, but both accurately reflect findings of the Alaska studies. With respect to the longitudinal study referenced above, the State of Alaska Department of Health and Social Services established a case definition as an increase in headache or an increase in at least two other symptoms (including nausea or vomiting, burning sensation of the nose or mouth, cough, dizziness, spaciness or disorientation, or eye irritation). Taxi drivers met the case definition more often than did health-care workers or university students (Beller et al., 1992 and Chandler and Middaugh, 1992). As part of the same study, CDC interviewed and collected blood samples from motorists, gas-station attendants, and mechanics living in Fairbanks, Alaska. It reported a higher prevalence of symptoms among workers interviewed during the oxygenated-fuel season than workers interviewed after the season. In addition, workers with MTBE blood levels in the upper quartile were more likely to report symptoms attributed to MTBE exposure, although the increases were statistically unstable due to the small sample size (CDC, 1993a, Moolenar, 1994).

In summarizing the retrospective study of outpatient insurance claims, the OSTP report concludes that in both Anchorage and Fairbanks there was an increase in visits for headaches in the winter of 1992-1993, when compared with the winter of 1990-1991, but not in the winter of 1991-1992, when the authors reported an epidemic of viral illness. The HEI report does not comment on this finding at all but states that the study indicates that "hospital admissions for respiratory ailments and asthma were stable over the 3-year period" (Gordian et al., 1995).

Neither the OSTP report nor the HEI report comments on the fact that given the uncertainties in exposure measurement in the cross-sectional studies, the majority of cases reported in both Fairbanks and Anchorage were in workers spending a longer amount of time in their automobiles. Likewise, in the longitudinal study conducted in Fairbanks, workers with higher MTBE exposures had an increase in symptom reporting. These results are internally consistent and suggest that workers occupationally exposed to MTBE are at greater risk for the development of acute health effects.

CONNECTICUT AND NEW YORK STUDIES

DATA REVIEWED IN THE HEI AND OSTP REPORTS. The HEI report reviews the results of a study conducted in Stamford, Connecticut, by White et al., 1995. The study was conducted during a time when the area was participating in the winter oxygenated-fuel program. A similar study of a comparison population was conducted in Albany, New York, an area where oxygenated fuel was not being used (CDC, 1993c). Although it is listed in the section entitled "Literature Cited," it is not clear if the HEI report also reviewed the CDC report on Stamford, Connecticut. The CDC report on Stamford, Connecticut, contains more information than does the paper published by White et al. in 1995.

The OSTP report reviews the White et al. (1995) paper on the Stamford, Connecticut, study, as well as the CDC report of that study. It also reviews the CDC report on the comparison population in Albany, New York.

In the CDC study conducted in Stamford, Connecticut, during the oxygenated-fuel season, the 11 people with the highest blood MTBE levels were more likely to report one or more key symptoms (including headache, irritated eyes, burning of the nose and throat, cough, dizziness, spaciness or disorientation, and nausea) than were persons with lower blood MTBE levels ($p < 0.05$) (CDC, 1993b, White et al., 1995). However, the overall prevalence of symptoms was similar in the population groups studied in Albany and Stamford. There was no difference in symptoms among those reporting use of premium fuels (with MTBE) to those using regular fuels (without MTBE) in Albany, New York (CDC, 1993c).

COMMITTEE CRITIQUE. Both the OSTP and HEI reports

accurately represent the Stamford, Connecticut, and Albany, New York, studies. Following an evaluation of both studies, the HEI report concludes that "although some symptoms were more prevalent in Stamford than in Albany, the pattern of elevation was not consistent." Thus, it discounts a relationship between MTBE exposure and symptoms, although this is not explicitly stated. The OSTP report notes that "qualitatively, the prevalence of the most common symptoms, such as headache and cough, occurring over the last month was not appreciably higher among men who worked around cars and gasoline in Stamford than men with similar occupations in Albany, where exposure to MTBE was generally much lower." This, too, in effect concludes that there were few data to support a relationship between MTBE exposure and the development of symptoms. The committee concludes that confounding due to exposure to gasoline itself and possibly temperature must be considered; however, despite confounding factors and instability due to small sample sizes, it is noted further by the committee that workers with the highest blood levels of MTBE in Stamford were more likely to report symptoms attributable to MTBE. This is consistent with symptom reporting among the occupationally exposed workers in the Alaska studies.

New Jersey Studies

DATA REVIEWED IN THE HEI AND OSTP REPORTS. The HEI report and the OSTP report review the New Jersey garage-worker study, which compared symptom reporting in state-employed garage workers during the winter oxygenated-fuel season in northern New Jersey and workers in southern New Jersey after the oxygenated-fuel season was over (Mohr et al., 1994). The HEI report also remarks on a paper by Fiedler et al. (1994) which compares symptom reporting, associated with driving and automobile refueling activities, in a small group of individuals diagnosed as having multiple chemical sensitivities and chronic fatigue syndrome and normal controls.

COMMITTEE CRITIQUE. Neither the HEI report nor the OSTP report notes that exposed workers who pumped gasoline more than 5 hours per day had an average increase in symptom score of 0.75 (from 2.26 preshift to 3.37 postshift). This is in contrast to an average improvement in symptom score of 0.45 (from 2.45 preshift to 2.00 postshift) at the end of the shift among low-exposed age-matched southern controls (total possible score, 28) (Mohr et al., 1994). Both reports note that this result was not statistically significant, but neither report comments on the consistency of this finding with results reported in occupationally exposed workers in Alaska and Stamford, Connecticut.

Neither the HEI nor the OSTP report comments on the abstract published by Mohr in which 107 older motorists (men 50-84 years of age) in New Jersey were surveyed for symptom reporting associated with length of time spent in their automobiles and gasoline refueling (Mohr et al., 1995). A few of the key symptoms were statistically significant associated with the number of hours spent in the automobile or the number of times per week that the automobile was refueled.

Fiedler et al. (1994) compared symptom reporting in individuals reported to be sensitive to very low levels of chemical exposures (multiple chemical sensitivities, MCS) to individuals with chronic fatigue syndrome (CFS) and normal controls. Based on the results from this study, HEI concludes that "further efforts to identify and characterize sensitive individuals are needed." The OSTP report states that groups studied thus far contained so few human subjects that comparisons are difficult to interpret. Neither the HEI report nor the OSTP report comment on the presentation made by Fiedler (1993) at the EPA Conference on MTBE and Other Oxygenates, although the HEI report lists this presentation in its

"Literature Cited" section. In this study, a small group of individuals, self-reported to the New Jersey Department of Health as sensitive to fuel containing MTBE, reported more symptoms associated with driving and refueling than did either the MCS or CFS controls. These individuals did not meet the author's MCS diagnostic criteria. The committee notes that the presentation made by Fiedler (1993) is the only scientific study done to date which may indicate that there are individuals sensitive to MTBE, and this study is limited in that it looked at only 5 or 6 individuals. In addition, the committee is surprised that this study is ignored in both reports which indicate further study on "sensitive" individuals is warranted. Thus, the committee questions the scientific basis of the recommendation by HEI that "further efforts to identify and characterize sensitive individuals are needed." It is the consensus of the committee that both the OSTP and HEI reports are way too premature in suggesting that almost all future research should be directed towards identifying and characterizing sensitive individuals and that occupationally exposed workers for whom there is consistent evidence of an increase in symptoms should be studied in more detail.

Wisconsin Studies

DATA REVIEWED IN THE HEI AND OSTP REPORTS. Both reports review a study in which the Wisconsin Department of Health and Social Services conducted a random-digit-dialing telephone survey of individuals regarding their concerns over the use of reformulated gasoline (Anderson et al., 1995a) and a followup study in which the same survey was administered to people who had already called Wisconsin government agencies with health concerns regarding reformulated gasoline (Anderson et al., 1995b).

COMMITTEE CRITIQUE. The HEI report concludes from the

first phase of the Wisconsin study that the "differences [in symptom reporting] are likely to be attributable to different levels of awareness about exposure to MTBE." This statement suggests that the HEI report supports the idea that symptom reporting in Wisconsin was mostly related to media attention. This is not entirely in keeping with comparisons made between Chicago and parts of Wisconsin outside the Milwaukee area, in which HEI reported that respondents in Chicago still reported a higher prevalence of eye irritation, headache, and sinus problems while pumping gas. In contrast, the OSTP report states, "the symptom prevalence was not elevated in Chicago compared with such prevalence in areas of Wisconsin where reformulated gasoline was not used." The committee reviewed the report by Anderson et al. (1995a). There are some differences in the way the symptoms are reported by region in relation to exposure in Tables 9, 10, 11, and 12. There is evidence in Table 12 that there was higher symptom-reporting prevalence in Chicago compared to areas in Wisconsin outside the Milwaukee area, and the conclusion drawn by the OSTP report is misleading.

As stated above, the same telephone-survey questionnaire was then administered to 1,339 persons who had called various Wisconsin government agencies with health concerns regarding reformulated gasoline. In this study, these contacts were slightly older and more likely to commute more than 1 h per day in a car than other Milwaukee respondents from the first phase of the study (Anderson et al., 1995b). The HEI report concludes that in this portion of the Wisconsin study, "exposure to RFG . . . was not a predictor of symptoms." The OSTP report concludes that the "health contacts also were more likely to have seen various news stories about MTBE than other Milwaukee residents." Both studies, although emphasizing different aspects of the Wisconsin survey, accurately reflect the findings.

OTHER STUDIES AND REPORTS REVIEWED IN THE HEI AND OSTP REPORTS

DATA REVIEWED IN THE HEI AND OSTP REPORTS. Other studies and reports reviewed by HEI include (1) the presentation by Livo which noted a decline in the total number of complaints received by the Colorado Department of Public Health and Environment over several years of winter oxygenated-fuel seasons (Livo, 1995); (2) a survey of physicians by the Missoula City-County Health Department in 1993 to determine whether illnesses had increased that winter (Missoula City-County Health Department, 1993); (3) a survey of workers in two oil refineries by Mehlman (1995); (4) a survey conducted by McCoy et al.(1995); and (5) a study surveying individuals from Anchorage, Alaska, for symptoms due to exposure to fuel oxygenated with ethanol (Egeland and Ingle, 1995).

Data reviewed by the OSTP report include (1) the presentation by Livo (1995); (2) a personal communication of health complaints due to MTBE from persons living in Colorado (Pat McCord, 1993), which was apparently not available to the HEI committee; (3) the report from the Missoula City-County Health Department (1993) and a personal communication from E. Leahy (1995) to obtain followup information from this community, which was apparently not available to the HEI committee; (4) a presentation by Raabe (1993) regarding complaints of workers from companies represented by the American Petroleum Institute; (5) a study by Medlin (1995) of workers represented by the Oil, Chemical and Atomic Workers Union; and (6) the study by Egeland and Ingle (1995). The OSTP report does not review the papers by Mehlman (1995) or McCoy et al. (1995).

COMMITTEE CRITIQUE. Although the CDC white paper states that both the HEI and OSTP reports reviewed the same studies, it is not entirely true. In addition, neither the HEI report nor the OSTP report draws major conclusions from the studies listed directly above in the section "Other Studies and Reports Reviewed in the HEI and OSTP Reports." The OSTP report draws on the differences between the Livo (1995) presentation and the personal communication from McCord (1993) to conclude that the "analyses of the number of complaints received by different health departments or trends in the number of complaints received over time are difficult to interpret." The HEI report quotes Livo (1995) in stating that overall there did not appear to be sufficient evidence to confirm a link between MTBE exposure and health complaints.

The HEI report concludes from the Missoula City-County Health Department (1993) study that "the information provided by the physicians did not indicate that an outbreak of illness occurred." The OSTP report draws on the personal communication from Leahy (1995) to conclude that during the following winter, ethanol rather than MTBE was being used and that "public concern over this issue essentially disappeared."

The HEI report acknowledges that Mehlman (1995) reports that some workers exposed to gasoline containing MTBE experienced some symptoms. The study by McCoy et al. (1995) demonstrated that distribution workers reported headache, dizziness, and nausea more often than service-station attendants, who in turn reported these symptoms more often than production workers. The HEI report notes that although the highest number of complaints was received in the winter during the oxygenated-fuel season, the actual manufacture and transport of MTBE had begun several months earlier. The OSTP report uses the presentation by Raabe (1993) to describe a part of the same study and to conclude that "the few complaints recorded each year suggest that this reporting system is not sensitive." The OSTP report concludes from the Medlin (1995) paper that oil, chemical, and atomic workers have reported many of the same symptoms that were reported by motorists in Alaska and other symptoms as well, including sinus problems, fatigue, and shortness of breath.

Both the HEI and OSTP reports failed to acknowledge that while largely anecdotal, these occupational studies are consistent with the studies of Alaska, Connecticut, and New Jersey, all of which reported that workers who are exposed to higher levels of MTBE may experience symptoms due to those exposures.

The OSTP report comments on the study by Egeland and Ingle (1995), which surveyed 100 adult residents in Anchorage, Alaska, for possible health problems due to the use of ethanol as an oxygenate. It concludes that the results of this study indicate a lower prevalence of symptoms than had been reported 2 years earlier, when MTBE was the oxygenate used in gasoline. The HEI report reviews this paper in its ethanol section and raises the concern that the questionnaire used focused on the same "key" symptoms first linked to MTBE exposure in Alaska for ethanol as well as MTBE exposure, although there was no clear-cut evidence that the symptoms associated with ethanol exposure would necessarily be the same as for MTBE. The HEI paper did note that a higher percentage of people noted an unusual odor while pumping ethanol-containing gasoline (vs. conventional gasoline) and more symptoms while driving than while refueling.

CONTROLLED-EXPOSURE STUDIES

EPA STUDY

DATA REVIEWED IN THE HEI AND OSTP REPORTS. Both reports review the paper by Prah et al. (1994).

COMMITTEE CRITIQUE. Prah et al. (1994) conducted a chamber study to determine if exposure to pure MTBE would elicit responses similar to those reported in Alaska following the addition of MTBE to winter fuels. Subjects were exposed to 1.4-ppm MTBE for 1 h. The only statistically significant response was that female participants rated the quality of clean air higher than air containing MTBE (Prah et al., 1994). Both reports conclude that the controlled-exposure studies failed to demonstrate either subjective symptoms or objective measures of eye or nose irritation and CNS dysfunction in healthy young adults exposed to MTBE vapor alone. The committee agrees with these conclusions, but emphasizes that this study does not replicate the setting of environmental and occupational exposures, which are chronic, frequently smaller, and to a mixture vs. a pure chemical. Thus, the relevance of these data to the interpretation of the acute health effects which have been reported in the field needs to be extended to include more environmentally relevant circumstances.

YALE STUDY

DATA REVIEWED IN THE HEI AND OSTP REPORTS. The HEI report reviews the paper by Cain et al. (1996). The OSTP report reviews a preprint copy of the same paper.

COMMITTEE CRITIQUE. To examine the effect of exposing humans to MTBE in combination with other volatile organic compounds (VOCs), Cain et al. (1996) performed a double-blind trial in which subjects were exposed sequentially to 1.7-ppm MTBE for 1 h on one day, uncontaminated air for 1 h 2 days later, and a 7.1-ppm mixture of VOCs used to simulate a gasoline mixture without benzene or MTBE for 1 h 2 days later. The subjects were able to accurately detect the odor of MTBE and the mixture of VOCs (Cain et al., 1996). The study also found that exposure to the mixture of VOCs led to an increase in polymorphonuclear neutrophilic cells. The HEI report concludes from this study that "a hydrocarbon mixture simulating gasoline did produce a minor inflammatory reaction in the upper airways as found on nasal lavage at a time delayed after the exposure. Thus, . . . one would expect vapor from gasoline containing MTBE to produce, at most, modest symptoms of irritation or inflammatory changes that do not differ substantially from those produced by gasoline vapor alone." The OSTP report notes that the finding of increased inflammatory cells in nasal lavage on the day following exposure was "consistent with previous work done at the EPA laboratories." The OSTP report cites work done by Koren et al. (1992) which was not reviewed by the HEI committee. The overall conclusions, however, were similar. The committee agrees with these conclusions but emphasizes that this study does not replicate the setting of environmental and occupational exposures, which are chronic, frequently smaller, and to a mixture versus a pure chemical. Thus, the relevance of these data to the interpretation of the acute health effects which have been reported in the field needs to be extended to include more environmentally relevant circumstances.

Swedish Studies

Data Reviewed in the HEI and OSTP Reports. Both reports review two presentations made by the Swedish National Institute of Working Life (Nihlen et al., 1994; Johanson et al., 1995).

Committee Critique. Nihlen et al. (1994) and Johanson et al. (1995) performed controlled-exposure experiments on healthy volunteers at higher concentrations (5, 25 and 50 ppm) than either EPA (Prah et al., 1994) or the Yale study (Cain et al., 1996). The HEI report notes that for all three studies "the blood levels of MTBE measured at the end of the exposure were much higher than the upper-quartile levels measured in workers and commuters in Stamford." The OSTP report notes that in the Swedish experiments there was a "marginally significant increase in . . . nasal swelling which was not concentration related." The committee agrees with these conclusions but emphasizes that this study does not replicate the setting of environmental and occupational exposures, which are chronic, frequently smaller, and to a mixture versus a pure chemical. Thus, the relevance of these data to the interpretation of the acute health effects which have been reported in the field needs to be extended to include more environmentally relevant circumstances.

ACUTE HUMAN HEALTH EFFECTS

EPIDEMIOLOGIC STUDIES

HEI REPORT. The HEI report states "formal epidemiologic investigations designed to test specific hypotheses concerning MTBE added to gasoline and specific human health or human comfort endpoints have not been undertaken." The committee disagrees with this statement. Several community studies were conducted in response to the public concerns that followed the widespread introduction of oxygenated fuel during the winter of 1992-1993. At least two of the studies had an open meeting where academicians, public-health officials, and industry representatives were invited to participate and provide input. At these meetings the goals of the studies were clearly presented, the dependent and independent variables defined, and the instrument used for outcomes measures extensively tested (Mohr et al., 1994; Fiedler et al., 1994; White et al., 1995; CDC, 1993b).

With respect to the epidemiologic studies, the HEI report concludes "some of these studies, for example, those in Alaska, New Jersey, and Wisconsin, were performed in response to outbreaks of complaints concerning symptoms after MTBE in gasoline had been introduced; in contrast, the studies in Connecticut and in New Jersey [sic] were carried out at times when outbreaks of complaints were not occurring." The committee notes that studies carried out in New Jersey were performed during the winter oxygenated-fuel season and during a period of time when increasing media attention was being placed on oxygenated fuels (Mohr et al., 1994; Fiedler et al., 1994). One study not referenced by either the HEI report or the OSTP report (Mohr et al., 1995) was carried out before much press coverage of oxygenated fuels had occurred.

The HEI report also notes that the epidemiological studies were limited in assessing the exposures of participants. It suggests that the symptoms reported in these studies were "mild and of short duration; no evidence of associated clinical morbidity has appeared although such outcomes were not specifically addressed." The HEI report attempts to summarize certain aspects of the epidemiologic studies in Tables 15-19. In Table 16 and 17 for the New Jersey study they use 30 day ever vs. never analyses of symptom reporting as "prevalence." This highlights the difficulty in comparing results from one cross-sectional study to the next, when the time frame used to define the prevalence of symptoms varied from 1 day to 3 months across the studies. They also list 835% [sic] of garage workers in southern New Jersey as having experienced skin irritation; this was not addressed in the New Jersey study.

The HEI report states that "the [epidemiological] studies described here do not provide definitive evidence for an association between exposure to MTBE and symptoms" and they go on to suggest a well-designed prospective study. It also suggests that "although more research is needed, these studies provide an indication that some individuals exposed to emissions from automotive gasoline containing MTBE may experience acute symptoms such as headache or eye and nose irritation." This suggestion fails to recognize consistencies in the data from many of the occupational studies of exposed workers. As stated previously, these studies indicate that workers exposed to greater concentrations of MTBE experience more symptoms. By failing to recognize the relationship between exposure and response and subsequently suggesting that "some individuals" may have problems (suggestive of a more sensitive subpopulation), they marginalize the problem. Even if a more sensitive subpopulation exists, the data presented indicate that more research is needed on exposed populations, rather than on sensitive subpopulations. The committee would like to point out that it is widely recognized among epidemiologists that most noninfectious diseases have multifactorial etiology. There are few if any deterministic relationships between risk factors and diseases, e.g., only 10% of heavy smokers get lung cancer in their lifetime (Rothman, 1986). However, it is generally regarded as an unproductive research goal to attempt through epidemiologic methods to identify those 10%; rather, epidemiology seeks to identify those factors that increase (or decrease) the population incidence or prevalence of disease.

OSTP REPORT. The OSTP report raises several concerns with respect to the epidemiological studies reviewed, including inadequate sample size, potential bias in the sample selection, inadequate or unreliable exposure information, and reliance on highly subjective measures of effect. With respect to the latter, the committee notes that acute-health-effects epidemiology must rely on these highly subjective measures of effect. Despite these issues, the OSTP report concludes that "taken together, these studies suggest that most people do not experience adverse health effects from MTBE in gasoline, but the studies cannot rule out the possibility that some people do experience more acute symptoms from exposure to oxygenated gasoline than to conventional gasoline." The report goes on to say "a causal association between acute health effects and exposure to MTBE or other oxygenates in gasoline in a relatively smaller proportion of persons has not been demonstrated but cannot be ruled out . . . [and] more definitive studies are needed."

Not unlike the HEI report, the OSTP report fails to recognize the consistency of the findings reported in the occupational studies of exposed workers. Despite their limitations, many of the studies and anecdotal reports suggest that workers exposed to higher levels of MTBE have an increased rate of symptom reporting. By suggesting that "a relatively smaller proportion of persons" may have problems with exposure, they ignore consistent findings in exposed workers and rule out the possibility that typical occupational exposures may pose a problem, whether or not a more sensitive general subpopulation does exist. However, the committee agrees with OSTP's discussion of the shortcomings of the available data: inadequate exposure assessment, insufficient sample size, potential for selection bias, and subjective outcome reporting.

CDC WHITE PAPER. The CDC white paper indicates that overall the statements made by the HEI and OSTP reports with regard to the epidemiological studies are consistent, even though the two reviews emphasized somewhat different aspects and limitations of the papers reviewed. No additional conclusions are put forth by the CDC white paper.

CONTROLLED-EXPOSURE STUDIES

HEI REPORT. The HEI report notes that while all three exposure studies failed to demonstrate subjective symptoms or objective measures of irritation or CNS dysfunction, these studies only exposed healthy young adults to MTBE alone for short periods and not as a component mixture of gasoline. In the executive summary, the Oxygenates Evaluation Committee identified a need for further human environmental-chamber studies to evaluate metabolism, symptoms, and neurotoxic effects in a more diverse population of individuals following exposure to MTBE and MTBE-gasoline mixtures.

OSTP REPORT. The OSTP report also states that controlled-exposure studies of healthy adults exposed to MTBE alone did not show an increase in symptoms or any notable adverse effects. The report also states that reported effects might be associated with other factors—such as low temperature, odor, or concurrent illness—that were not reproduced in the controlled-exposure studies. This report concludes that "these findings do not rule out the possibility that a subpopulation of people in the general population may be especially sensitive to MTBE alone or in gasoline." This report also recommends further studies among volunteers with self-described sensitivity to oxygenated fuels or combustive and evaporative emissions.

CDC WHITE PAPER. The CDC white paper reaches the conclusion that both reports "recommended that additional experimental studies of the mixture of MTBE and gasoline or similar hydrocarbon mixture be undertaken on potentially sensitive subjects."

CONCLUSIONS

- The committee believes that while the epidemiologic data currently available do not establish a causal relationship between exposure to gasoline containing MTBE and the development of symptoms, the studies do indicate that some people have experienced acute symptoms associated with exposure to gasoline containing MTBE. Limitations of the studies make it difficult to rule out the possibility of an association between exposure to gasoline containing MTBE and acute symptoms, and the studies contain enough suggestion of a dose-response effect to motivate further investigation.
- The health effects investigated thus far have all been acute symptoms. No data are yet available to link MTBE exposure to measurable adverse public-health effects, such as increased visits to

health-care providers or increased days of missed work or to the development of clinical acute or chronic disease.

▪ The data indicate that there is enough consistency among various studies to suggest that the levels of exposure to gasoline containing MTBE in certain occupational settings are associated with increased rates of symptom reporting.

▪ There are not sufficient data to indicate that the reported acute health effects are confined to a sensitive subpopulation. Therefore, it seems inadvisable to focus research on such subpopulations.

▪ The committee believes that there should be coordination between clinical observations, epidemiology studies, and exposure-chamber experiments regarding acute human health effects and exposure to oxygenated fuels.

RESEARCH NEEDS

EPIDEMIOLOGY

One or more analytical epidemiologic studies examining the association between exposure to MTBE resulting from winter oxygenated-fuel programs and acute health effects should be conducted.

Such studies should have the following features:

- Exposure to MTBE should be quantitatively assessed on an individual level through the determination of MTBE concentrations in blood and breathing zones (TBA concentrations can be considered in place of MTBE, if the validity of TBA as a surrogate for MTBE is established in a pilot study of sufficient sample size).
- Symptom prevalence should be assessed through questionnaires. To the extent possible, the reliability and validity of these

questionnaires should be determined in a pilot study. Validity can be established, at least in part, through coordination with chamber studies (see recommendation below), where symptoms are provoked and objective measurements are taken simultaneously.

• Outcomes which can be assessed through objective measures should be included whenever possible, e.g., the inclusion of CNS function measures using the protocol of Baker et al. (1985) to provide assessment of "key" neurological symptoms—such as dizziness, light-headedness and sleepiness, and ocular hyperemia-redness—and possibly immunological measurements from blood samples, because some data (Anderson, et al. 1995b) suggested a higher prevalence of MTBE-associated symptoms among those with physician-diagnosed allergies.

• High-quality data on potential confounders should be collected at the level of the individual and should include sex, race, age, educational level, occupation, the Barsky (1990) Amplification Scale, current use of medications, features of any automobiles driven (including vehicle year and status of antipollutant technology), type of gasoline used, amount of time spent in automobiles and in traffic, weather conditions at the time of measurements (including temperature, humidity, and air pressure), and concurrent exposures in both the blood and breathing zone (including benzene, toluene, formaldehyde, acetaldehyde, carbon monoxide, and xylene).

• Power and sample-size calculations should be done carefully to ensure sufficient statistical power to detect health effects corresponding to a 50% increase or more (i.e., relative risk of 1.5). These calculations should consider the high correlation of simultaneous exposure to numerous chemicals with possible health effects.

• Cross-sectional designs as well as "longitudinal designs" (preshift vs. postshift or beginning of day vs. end of day) should be considered, or perhaps a hybrid of these. These designs are relatively quick and inexpensive to conduct. Optimally, a more expensive, time-consuming design to consider is a longitudinal study in which a sample of individuals are enrolled 1-2 months prior to the commencement of the winter oxygenated-fuel program and monitored at regular intervals subsequent to the commencement of the program and until its termination in the spring. Symptom diaries can be kept between monitoring visits. This design would be able to provide information on the time course between exposures and possible outcomes, as well as the possible development of sensitivity and/or tolerance to exposures over time. Since individuals are used as their own controls, confounding by many individual factors is eliminated.

- To improve cost effectiveness of any study (studies), sophisticated modern designs in which detailed confounder, exposure, and outcome data are obtained in a subsample should be considered (two-stage designs) (Cain and Breslow, 1988; Tosteson, TD and Ware, JH, 1990). Since standard statistical software is generally not applicable when these designs are used, the assistance of a statistician with experience in the design and analysis of such studies will be required.
- To ensure generalizability, study participants should include adequate representation of both environmental and occupational exposure levels.
- Efforts to maximize participation rates of sampled individuals should be considered and should not exclude financial honoraria for participation; in addition, partnership with the gasoline industry should be considered, to facilitate access to exposed individuals and workers at filling stations and garages.

CHAMBER STUDIES

The reports reviewed by the committee noted that the "stimulus" of the effect requires further study. This would include the effects of any fuel oxygenate, such as MTBE, alone, the effects of an oxygenate in a background of gasoline vapors, and the effects of vapors from combustion of oxygenated fuel. In addition to the stimulus, the *system or modality* of an effect (e.g., irritation of the eyes and nausea), the appropriate *response* (e.g., rating of severity), and any *predisposing factors* (e.g., age) should be investigated.

STIMULUS. Within the context of chamber studies, the committee endorses the need a) to study MTBE or other oxygenates against backgrounds of gasoline or combustion products (at open-loop and closed-loop conditions of operation of an engine) as deemed ethical, b) to vary the temperature of exposures in chambers in order to simulate cold-weather vs. warm-weather conditions, and c) to repeat exposures over time in order to simulate real-world circumstances. A series of chamber studies could not only help in development of a structured protocol for evaluation of acute human health effects of MTBE, but also of other oxygenates and of VOCs in general. (Protocols developed to explore acute effects of oxygenates might prove to be useful to explore effects associated with sick building syndrome and multiple chemical sensitivity as well.)

If an oxygenate is in use in a geographic area, it should be possible to recruit people from that area and from suitable control areas to record their symptoms or observations as they occur in everyday life. (Involvement of designated clinicians to examine any symptomatic individuals in such a group could prove quite valuable in this context.) Data collection and analysis would need to be sensitive to the variables of concentration, time, other situational factors, and susceptibility of the host. Observations of this nature would not prove cause and effect so much as they would narrow the range of possibilities for more controlled study. For example, they would help to set up uniform reporting requirements and thereby reduce the likelihood that symptoms would go unreported. They would also reveal, via accompanying monitoring, how symptoms may vary with time and level of exposure; i.e., symptoms might lag behind exposures by some period of time or occur more frequently when a person is fatigued or when a person's ventilation is above average, as during physical labor. Since exposures in chambers cannot reproduce every circumstance of life, these prior observations could help to focus chamber studies on the most relevant circumstances.

Studies in chambers can in principle simulate many field exposures. If, for example, one learned from field observations that brief, high exposures seemed to trigger symptoms that persist for a while whereas longer, low exposures did not, then this could productively direct research in chambers. Merely measuring every conceivably relevant symptom using arbitrarily decided durations or duty cycles will detract from productive hypothesis-testing. The design of chamber studies should be informed by epidemiological studies and any clinical observations mentioned above and in turn, the chamber studies should entail development of techniques that may prove useful in epidemiological studies and perhaps in clinical objective validation of symptoms.

MODALITY. Symptoms reported in the field often involve irritation of the eyes or nose. In laboratory experiments, irritation in the nose or eyes can be induced without any systemic exposure. For example, induction of local irritation in the nose can occur via voluntary velopharyngeal closure (i.e., closing off the nasal cavities from the pharyngeal area) and restriction of oral breathing. Stimulation of the eyes requires no inhalation. Measurements of thresholds for irritation in brief or longer-term exposures could reveal the distribution of sensitivity in normal individuals. Investigation of a sample of appropriate size and composition would in essence allow a definition of hypersensitivity.

At present, the most feasible assays for human sensory irritation are psychophysical. These should entail exposures for durations commensurate with the time constants of chemosensory functioning. Because irritation may take minutes or even tens of minutes to develop, exposures should last a commensurate length of time.

From the standpoint of objectivity, psychophysical assays of irritation should entail blinded exposures, i.e., subjects should not be able to determine the presence or the level of a stimulus via smell. One possibility for "blinding" involves "separation" of the eyes from the upper airways in such a way as to remove olfactory cues. Blinding during nasal stimulation poses a more difficult challenge. Odor masking offers some possibilities, but requires the introduction of another chemical. Another possibility for blinding would involve use of anosmic subjects. In such subjects, velopharyngeal closure would, as in normal individuals, avoid systemic exposure. Use of such subjects would need to include tests for their comparability with normal individuals. (In major cities, it is not difficult to recruit a dozen or more persons with verifiable anosmia.)

There has been little study of how the eyes respond to chemical exposure that do not result in frank inflammation. For example, it is unknown whether susceptibility to irritation varies systematically during the day or as a function of the performance of particular tasks. Because of the absence of this background information, putative correlates of irritation, such as tear-film breakup and evaluation of scoring of the conjunctiva, both of which have been used in chamber studies, remain poorly verified with respect to reactions caused by exogenous chemicals. The same holds true for objective indexes of nasal irritation. All the various potential objective correlates of what we can term surface symptoms require further validation, which can only be done if there is a deliberate attempt to precipitate reactions. Such efforts to correlate symptoms with objective indexes may help determine requisite statistical power for measuring what might be meaningful changes in symptoms or their correlates.

Aside from chemosensory measurements, which are necessary to specify the level at which a chemical may become bothersome, other measurements should be made with some hope of understanding the modality through which they operate. Field data have provided only weak specification of the symptoms allegedly caused by MTBE. For even a chance of appropriate specification of symptoms, data need to be collected prospectively. The possibility that nonspecific symptoms attributed to exposure to MTBE may have resulted from other causes requires careful attention. Not only is it a waste of time and money to study effects that may not be real, but such studies lead to indeterminacy and continued unresolved concerns.

Without understanding of why and when symptoms occur, there can be no true honesty in dealing with complaints. The current level of understanding allows no probative answers; accordingly, decisions regarding exposures to oxygenates may devolve to nonscientific considerations.

RESPONSE. Following characterization of stimulus-driven symptoms and their correlates, attention can presumably focus on the most important symptom and its objective correlate. This symptom should be studied in hypothesis-driven experiments. If the symptom were headache, for example, research would need to (1) establish the minimal conditions of exposure for it to occur in the typical case and (2) examine its correlates, e.g., blood flow and accompanying blood levels of oxygenate and/or gasoline. It is important to emphasize that the key symptom should be studied in order to be understood, not merely cataloged. If a second symptom seems relevant from field data, it too should receive attention.

A surface potential from the respiratory mucosa (i.e., the negative mucosal potential), if proved comparable with human psychophysical results on irritation, could provide an objective index even for subjects whose sense of smell is allowed to operate normally during exposure.

Separation of the surface effects of MTBE, gasoline, or any candidate oxygenate from systemic effects could in principle simplify the study of acute reactions. For example, a study involving measurements of surface effects in anosmic persons in the manner described above can also include determining inhaled concentrations of MTBE that would give measurable blood concentrations. If elevated blood levels are not associated with sensitivity to surface exposure, then surface effects and systemic effects could perhaps be studied independently. Such an outcome may ultimately help determine whether surface effects "drive" other symptoms; e.g., irritated eyes might drive feelings of fatigue or listlessness, which might lead to reports of "spaciness."

Along the lines of separating surface from systemic effects, alternative routes of administration of the chemical of interest might prove worthy of consideration. Ingestion or infusion might replace inhalation as a way to investigate systemic symptoms or signs of impairment in humans. These alternative routes may permit double blinding more readily, though this would need to be demonstrated. (In theory, exhaled oxygenates could thwart blinding.) The time-course of blood concentrations of the relevant agent introduced by an alternate route would need to mimic that obtained via inhalation. Insofar as the principal systemic effects were neuropsychological in nature, the use of a positive control of ethanol would seem merited.

PREDISPOSING FACTORS. Field experience indicates that not every person will experience symptoms of exposure to oxygenates under common environmental conditions of exposure. Predisposing factors, insofar as they are demographic pr medical characteristics — such as sex, allergies, and history of exposure — could be studied in hypothesis-testing.

With respect to the recommendation for "additional controlled exposures of people with self-described sensitivity to oxygenated fuels" (OSTP report, p. 59), some people may indeed have greater sensitivity than others to MTBE, but it seems premature to focus research on such individuals without objective criteria to suggest the existence of such a group. Nevertheless, some pilot testing toward development of criteria would not seem out of place. Although the majority of people exposed to MTBE via oxygenated fuel may voice no complaints, this does not mean that they fail to experience many of the same reactions as those who do complain or would complain under the same circumstances of exposure. There is presumably too little known about the interaction of circumstances of exposure and susceptibility to make the case for the existence of a sensitive group, as opposed to particular circumstances that precipitate complaints. Conceivably, persons who complain have experienced special circumstances (e.g., fatigue at the point of exposure or airway disease). The tendency to attribute the exposures in a chamber to a hypothetically sensitive group could prove unproductive or possibly even counterproductive in an environment of limited resources. The reports acknowledge the severe limitations of previous chamber studies, which would seem to argue strongly for further studies of normal subjects under various conditions of exposure (e.g., longer durations, use of mixtures of MTBE and gasoline) and of subjects with objective indications of risk (e.g., asthma).

In summary, the chemosensory properties of potential oxygenates form important parameters for detection and possible annoyance effects and should be delineated expeditiously. Symptoms associated with real-world exposures to agents should be characterized from prospective field data in order to be used in more formal studies. Such studies should offer insight into how a symptom occurs and should entail focused hypothesis-testing. Re-creation of the symptom in a chamber should, if possible, be accompanied by an objective correlate to verify its presence and perhaps to add insight into its mechanism. Field data could offer insights into predisposing factors for symptoms, with followup in formal studies. This work will undoubtedly yield both methodological and substantive insights as studies are done. The methods used in human chamber studies may seem as routine as that used in toxicological screening, but they are not. There is a great deal to be learned about how to study such problems as those posed by MTBE. Researchers in this area need to refine their techniques with actual experience. This does not argue for an open account, but argues for acknowledgment that as VOCs are considered as potential oxygenates, research on their potential acute effects should begin with greater knowledge of how to screen for such effects than is now available.

ODOR STUDIES. For some oxygenates, the olfactory threshold is already known; for others, the data should be gathered.

There may be merit in conducting further studies of odor detection of oxygenated vs. conventional gasoline. One question that often arises is whether persons become more sensitive over time to the presence of oxygenates. Within this context, it may be advantageous to determine if people become primed by the odor of oxygenates. Perceptual priming has been the subject of hundreds of studies of which only few focused on olfaction.

In light of the high suggestibility toward false alarms in olfaction, it would be of interest if subjects show more likelihood of perceiving oxygenate when it is not present.

Insofar as odor of oxygenates plays a role in exacerbation of asthma in asthmatic patients, nasal exposures to odor independently of systemic exposure may help to determine whether the exacerbation might be "caused" by the odor.

ECOLOGICAL STUDIES

Due to the difficulties (e.g., high cost, long followup time, and large required sample size) of designing an analytical epidemiologic study in which individual data are linked to cancer and other chronic morbidity outcomes, attention must be turned to ecologic designs, despite their widely acknowledged deficiencies. Death-certificate data are collected by all states, making it possible to link exposures with cancer, cardiovascular, and respiratory mortality, by ICD code. In addition, nearly all states have cancer registries, making it possible to calculate cancer incidence rates. Geocoding by the unit of environmental sampling should be considered as a method for controlling for bias due to differences in the distribution of other determinants of the diseases under study. However, sufficient exposure data at the community level which could be linked to routine mortality and cancer incidence data are not presently available, so such studies are impossible today. Thus, the committee does not recommend that ecological studies be undertaken at the present time, but recommends that the most important pollutants be identified and environmental monitoring data, such as those discussed in chapters 2 and 3, begin to be collected, so such ecological studies can be conducted in the future, perhaps in 10 years or so. Measurement techniques should be developed for pollutants for which adequate techniques are not available.

REPRODUCTIVE AND DEVELOPMENTAL EFFECTS

DATA REVIEWED IN THE HEI AND OSTP REPORTS

The HEI report reviews the data available on the reproductive and developmental effects of MTBE and ethanol. A number of animal studies have been conducted on the potential reproductive and developmental effects of MTBE. Maternal toxicity and effects on reproductive indexes have been observed at exposure concentrations of 3,000-8,000 ppm. In offspring, developmental effects have also been observed at these concentrations. No maternal or developmental effects have been observed at lower concentrations (300-

400 ppm). Actual human exposures are expected to be lower than these no-observed-effect levels by a factor of 1,000 or 10,000.

Ethanol is a known potent developmental toxicant in humans. However, it is not anticipated that exposures to ethanol from use of ethanol as an oxygenate will result in a substantial deviation in blood ethanol from endogenous levels.

The OSTP report reaches essentially the same conclusions with regard to the potential for reproductive and developmental effects in humans.

COMMITTEE SUMMARY

The committee agrees with both the HEI and OSTP reports that adverse reproductive and developmental effects are not expected to result from MTBE exposure at the levels at which most people would be exposed.

RESEARCH NEEDS

The HEI report does not give high priority to research on reproductive or developmental effects. The committee feels this is reasonable. However, appropriate studies should be conducted on other proposed oxygenates before they are extensively introduced in the U.S. gasoline supply.

LONG-TERM HEALTH EFFECTS

HUMAN

Both the HEI and OSTP reports failed to identify any human

studies on the potential long-term health effects of MTBE or other oxygenated fuels. Chronic health effects due to the ingestion of ethanol have been well studied and are not addressed in this review.

ANIMALS

DATA REVIEWED IN THE HEI AND OSTP REPORTS

As noted in the CDC white paper, while there are many similarities between the HEI and OSTP reports regarding the carcinogenicity of MTBE, there are also notable differences. The HEI report gives a more comprehensive review of the various animal studies than does the OSTP report. Neither the HEI nor the OSTP report evaluates the potential of other ether oxygenates, because no long-term carcinogenicity studies have been conducted.

COMMITTEE CRITIQUE

ORAL ADMINISTRATION IN RATS (BELPOGGI ET AL., 1995). The HEI and OSTP reports accurately describe the results of the study, but both fail to note the deficiencies in study design, conduct, and reporting. For example, decreases in survival of 15% (at 250 mg/kg) to more than 20% (at 1,000 mg/kg) in females were noted as early as 9-12 months. That there was no concurrent effect on bodyweight gain suggests that the deaths were directly attributable to the toxic effects of MTBE, and therefore that both exposure levels probably exceeded the definition of a maximum tolerated dose (MTD). Survival in males was also decreased but not until late in the study.

There is no information on the cause of death in the early-death females which would help in the interpretation of the study. In addition, no rationale is given for the selection of dose levels or the highly unusual dosing regimen, i.e., dose on Monday and Tuesday, no exposure on Wednesday, then dose on Thursday and Friday.

As noted in both the HEI and OSTP reports, a dose-related increased incidence of lymphomas and leukemias (combined) was observed in female rats, but neither report mentioned or highlighted the fact that it was found only at doses which clearly decreased survival and that no such increase in tumors was observed in male rats where survival was not so impacted. Also, there was no description of the morphologic criteria for these two entities.

The other reported increase in tumor response related to MTBE exposure was interstitial-cell (Leydig's cell) tumors of the testes at 1,000 mg/kg, but not 250 mg/kg. Neither the HEI nor the OSTP report notes that the Belpoggi et al. (1995) study fails to give any description of the lesions in terms of size or the criteria used for their diagnosis, which is extremely important for these types of neoplasms, which have a morphologic continuum from hyperplasia to neoplasia. Also, unlike the lymphoma-leukemia situation, where possible "preneoplastic" lesions of the same cell type are described and tabulated, no such Leydig's cell proliferative lesions are mentioned. A concordant increase in hyperplastic lesions would add evidence to the observations. Neither report takes note of the fact that the high-dose male rats showed increased survival after 88 weeks (compared to the middle-dose and controls) and therefore were at greater risk for development of these late-appearing neoplasms (first tumor was found at 96 weeks). It is not clear if the statistical analysis adjusted for this.

Finally, because of the importance of this study for eventual use in risk assessment, the superficial reporting of the data, and the nature of the observed lesions, the committee feels strongly that an independent in-depth review of the data, especially the pathology (microscopic slides) of the critical lesions, is warranted (as was done with the inhalation studies) before the data are used for risk assessment.

INHALATION EXPOSURE IN RATS. Both the HEI and OSTP reports state that the increased incidence of renal adenomas in male rats was related to MTBE exposure and speculate that it may be due to the metabolite TBA. Both reports discuss in some depth the possibility of α_{2u}-globulin's being responsible for the kidney tumors but conclude that this is probably not the case. The committee, after hearing about and evaluating studies (Prescott-Mathews et al., 1996; Poet et al., 1996) conducted recently at the Chemical Industry Institute of Toxicology, feels that α_{2u}-globulin may, in fact, be involved in the causation of these neoplasms. It appears that this research has fulfilled the EPA criteria of causation in this regard. In any event, both reports need to revisit this issue in light of these new findings.

INHALATION EXPOSURE IN MICE. Both the HEI and OSTP reports are in agreement with regard to the principal finding in this study, i.e. the induction of liver tumors. There was a mild but statistically significant increase in benign (only) tumors in females at the high dose (8,000 ppm), which was also statistically significant (compared to controls) when benign and malignant tumors were combined. In contrast, while there was an increase in carcinomas in male mice at 8,000 ppm (only), there was no statistically significant increase when they were combined with adenomas — the more appropriate method of analysis.

The question raised in both reports was whether the study was of sufficient length (18 months) to support conclusions concerning the carcinogenic potential. Both reports suggest that the study should have gone to 24 months, as is routine in the National Toxicology Program (NTP). However, the committee felt that both reports fail to note that CD-1 mice (used in the inhalation study) do not live as long (on average) as the standard NTP mouse (B_6C_3F1) and that therefore the CD-1 mouse study may not be adequate for risk-assessment purposes. Moreover, the quantitative potency estimate derived from this study included an upward adjustment of risk to account for the "less than lifetime exposure."

The committee agreed with the discussion in the HEI report concerning the possible mechanisms for induction of the liver tumors and also agreed that nongenotoxic hormonally related mechanisms were the most plausible explanation. Thus, the committee feels that it is inappropriate to combine the male and female tumor responses for determining "maximum likelihood estimates" and "upper confidence limits," as was done in the OSTP report.

COMPARISON OF ANIMAL STUDIES. Neither the HEI nor the OSTP report discusses comprehensively the long-term animal studies in their totality, i.e., a weight-of-evidence approach. While the reports note that MTBE is a multispecies, multisite, and multisex animal carcinogen, they fail to take note of the inconsistencies in this regard as follows.

LYMPHOMAS AND LEUKEMIAS. There was a reported increase in these tumors in female SD rats exposed via gavage at 1,000 mg/kg but not at 250 mg/kg, nor in males at either dose. In contrast, neither sex of F344 showed this effect even at an air concentration at 8,000 ppm, which corresponds to a dose at least 4 times greater than the largest gavage dose, that was clearly toxic. This incongruity is especially noteworthy in light of the fact that this strain of rat (F344) is highly prone to the induction of leukemia, which is in fact an important cause of death in older F344 rats.

RENAL TUBULAR-CELL TUMORS. While the inhalation study in F344 rats showed an increase in these tumors in males (only) at 3,000 and 8,000 ppm, no such effect was found in the SD rat. One has to question why this is so in light of the fact that both strains of rats are susceptible to α_{2u}-globulin nephropathy and resulting tumors. While the routes of exposure are different, this is a systemic effect which should be present with either exposure regimen.

One possible explanation (which should be addressed in the OSTP report) is that the highest gavage dose is still approximately one-third less than the lowest inhalation dose. However, because of the bolus nature of the gavage dose, it would be expected that peak blood levels would actually be greater in the gavage exposure than in the lowest inhalation exposure, "3,000 ppm," in which these tumors were observed. However, the situation is complex because the metabolite of MTBE, TBA, may be responsible for the effects. In any case the OSTP report should discuss this issue in some depth in an attempt to explain this apparent discrepancy in results.

LEYDIG'S CELL TUMORS. Both the inhalation and gavage studies report an increase in Leydig's cell tumors of the testes, although the increase is not as impressive in the inhalation study, probably due to the high spontaneous rate in this strain (F344) of rat. However, this response is inconsistent with the TBA study conducted by the NTP in the F344 rat, where no such increase was observed. It is unclear why this should be the case in light of the proposed (HEI report) mechanism of tumor induction.

Such inconsistencies are rather unusual in rodent bioassays, especially when it is clear that the dose levels are at or above the MTD, as is the case in these studies. At a minimum, these inconsistencies need to be investigated in some depth before the animal data are used for risk assessment. To resolve this issue, the committee suggests that the Belpoggi et al. (1995) pathology slides be reviewed by a group of independent pathologists, as was done with the inhalation studies of MTBE and TBA, to verify the findings and resolve these inconsistencies.

GENOTOXICITY. The committee agrees with the HEI and OSTP reports, although it was noted that both are incomplete, i.e., each contains studies that are missing in the other report. The committee feels they should be reconciled. As noted in the HEI report, little information on the genotoxicity of the other oxygenates is available, other than ethanol.

MECHANISMS OF TOXICITY. The HEI report gives much more attention to the issue of possible mechanisms of carcinogenicity than the OSTP report. In fact, this is an important component of the report and should be considered before conducting any risk assessment.

CONCLUSIONS

- Because of the inconsistencies and unsolved questions with regard to the animal-carcinogenesis studies, cancer-potency estimates of MTBE as proposed in the OSTP report should be considered carefully. The committee feels that the male rat kidney-tumor data probably should not be used for this purpose in light of the new information on its probable causation, i.e., α_{2u}-globulin nephropathy, which is thought to be unique to the male rat and not relevant to humans.
- The use of the lymphoma and leukemia data should also be questioned until a thorough review of this study, including an objective third-party review of the pathology, is accomplished.
- The most reliable data available for risk-assessment purposes are on the induction of benign liver tumors in female mice exposed to 8,000 ppm MTBE via inhalation. Although it should be recognized that this amounts to extremely weak evidence of carcinogenicity, it cannot be discounted.

RESEARCH NEEDS

If other oxygenates become or are expected to be used in formulated gasoline to which large numbers of humans are exposed, in-depth chronic-toxicity and carcinogenicity studies in animals should be conducted (before the introduction of other oxygenates).

6

Potential Health Effects of Other Pollutants

The HEI Oxygenates Evaluation Committee report identified three main areas of concern from MTBE exposure—including acute symptoms associated with short-term exposure, potential neurotoxicity, and potential carcinogenicity—but because of limited data little emphasis was given to other pollutants that may result from use of oxygenated fuels. Although the goal of the winter oxygenated-fuel program is to reduce ambient carbon monoxide (CO) pollution to protect public health, particularly patients with cardiovascular disease, who are most susceptible to the adverse effects of CO, at least two factors may limit the overall effectiveness of the program, including insufficient lowering of ambient CO with these fuels and an increase in other pollutants because of fuel oxygenates. The main conclusions in the HEI report regarding these two factors are as follows: "the health benefit of reducing CO is uncertain because of severe limitations in the information about the number of sensitive individuals with coronary artery disease, their personal exposures to CO, and their activity patterns" (page 101); "the short-term symptoms reported with using MTBE are not unlike those sometimes reported after exposure to gasoline vapors or motor vehicle emissions from conventional gasoline" (page 101); and "our current understanding of the human effects from various carcinogens in motor vehicle emissions (including MTBE and those air toxics that have increased or decreased levels when oxygenates are used) is not sufficient to make confident predictions about whether using MTBE will cause any overall increase or decrease in the total carcinogenicity of emissions compared with using conventional gasoline. However, a substantial increase in carcinogenic risk from using gasoline containing MTBE or ethanol is not expected" (page 101).

COMMITTEE CRITIQUE

The literature review and conclusions of the HEI report concerning oxygenated fuels and their effect on other pollutants—including CO, benzene, 1,3-butadiene, formaldehyde, and acetaldehyde—provide a qualitative assessment of the limited information that is available. However, several statements require further detail and/or clarification. Because several data sources are available, the phrase "severe limitations in the information about the number of sensitive individuals with coronary artery disease" seems to overstate the problem. More emphasis needs to be placed on the emerging human data on air toxics and cancer (HEI report, page 94) in order to determine which animal models are relevant for human cancer risk assessment. A more comprehensive evaluation of the overall effects of oxygenated fuels on air quality is a major objective of a review being conducted by OSTP (HEI report, page 91). In addition, while the HEI report identified large gaps in our knowledge about the relationship between oxygenated fuels and their potential effects on ambient air pollution and health effects, specific research recommendations were not provided.

The OSTP report *Interagency Assessment of Potential Health Risks Associated with Oxygenated Gasoline* did not address changes in air quality or potential health effects associated with other pollutants that may result from oxygenated fuels. These factors will be considered in the second phase of the review that is anticipated to be completed in mid-1996 (page iii). Thus, the committee agrees with the conclusion stated in the OSTP report (page 64) that the "continued use of oxygenated gasoline will require a continuing evaluation of potential health effects (short and long term) of the oxygenates, their metabolites, and any emission or atmospheric degradation products . . . such as t-butyl formate."

In addition, the committee agrees with statements endorsed in the executive summary of the HEI report (page 4) concerning gaps in our knowledge about the health effects of MTBE and other pollutants from the use of oxygenated fuels. "In addition to its conclusions about possible health effects, the Oxygenates Evaluation Committee noted a general lesson to be learned from introducing oxygenates to the general public. Although it is not possible to have complete information about a substance before it is used, the diverse experiences after introducing oxygenated fuels argue strongly that any future new use of a substance should (1) be preceded by a sufficiently comprehensive research and testing program (including mechanistic and human studies), and (2) be accompanied by rigorous exposure assessment and epidemiologic studies."

CONCLUSIONS

Although the goal of the winter oxygenated-fuels program is to reduce ambient CO levels to protect public health, particularly among patients with cardiovascular disease, data are not available to evaluate the effectiveness of the program in protecting human health.

RESEARCH NEEDS

Because the goal of the winter oxygenated-fuel program is to lower ambient CO with the specific aim of reducing carboxy-hemoglobin and risk for exacerbation of cardiovascular disease, a fundamental question that has to be addressed is how effective the program is in accomplishing these outcomes. While this question will be considered in a subsequent OSTP report (OSTP report, page iii), research recommendations and priorities about other pollutants are fundamental for a comprehensive evaluation of the oxygenated-fuel program. Specific research recommendations are presented below.

CARBON MONOXIDE

- Epidemiologic studies to determine the effect of oxygenated fuels on population distributions of personal exposures to CO and/or of carboxyhemoglobin levels among patients with cardiovascular diseases.
- Epidemiologic studies of carboxyhemoglobin levels and risk for cardiovascular morbidity and mortality among patients with cardiovascular diseases.

AIR TOXICS AND SHORT-TERM HEALTH EFFECTS

- Epidemiologic studies to determine exposures to air toxics that may produce short-term health effects from combustion of oxygenated fuels and conventional gasoline.

- Experimental studies to examine the potential irritant effects of aldehydes and other toxics at levels likely to be found from the combustion of oxygenated fuels and conventional gasoline under real-life conditions (e.g., low temperature).

OTHERS

- Cost-effectiveness study of the winter oxygenated-fuel program using data from studies described in the recommendations above.
- Risk assessment using data from the recommendations described above.

7

Risk Assessment

This chapter contains a synopsis of the approaches to the assessment of human health risks adopted in the HEI and OSTP reports. It also contains an evaluation of the scopes of the two reports, the data and methods used in each, and recommendations for further development of the risk-assessment portion of the federal government's comprehensive evaluation of oxygenated fuels now being undertaken under the aegis of the NSTC's Committee on Environmental and Natural Resources. The final part of this chapter contains recommendations for research efforts that could improve the scientific basis of risk assessment.

SYNOPSIS OF THE TWO REPORTS

HEI Report

Scope

The HEI report is broad in scope. It attempts to describe comprehensively the human health risks associated with the use of gasoline containing MTBE or ethanol; other oxygenates are discussed, but, because health-effects and human-exposure data for these compounds are extremely limited, risk estimates for them could not be developed.

With respect to fuels containing MTBE or ethanol, the HEI report notes the absence of toxicity data on the actual fuels[1], and resorts to data available on selected chemicals, including the two oxygenates. Other chemicals from the fuels (both evaporative and combustion emissions) considered to be potentially important are formaldehyde, acetaldehyde, benzene, and butadiene. These later substances are, in effect, taken as at least partially representative of the potential carcinogenic risks associated with the use of gasoline. The HEI report qualitatively evaluates how the health risks of each of these selected compounds might change when the oxygenates are introduced. These risk comparisons are shown in Table 7.1, taken from the HEI report. Note that the HEI report also considered the potential health benefits associated with the use of oxygenates, represented by their effects on CO exposures.

DATA, METHODOLOGY, AND CONCLUSIONS

The HEI report presents a comprehensive review of all data relevant to the risk questions being posed. As is appropriate for human health risk assessment, emphasis is placed on documented health effects, demonstrated in human and/or animal studies. Data gaps are noted, and judgments are made regarding risks about which some conclusions could be reached (i.e., those in Table 7.1).

---

[1]Although there are animal-carcinogenicity data on unoxygenated gasoline, these are appropriately considered by HEI to be inadequate for evaluation of human risk, because the test animals were exposed to totally vaporized gasoline, a mixture to which humans are not exposed.

Table 7.1 Potential health effects of low-level exposure to various pollutants and the projected direction of change in exposure levels for each pollutant when oxygenated fuel or reformulated gasoline containing MTBE or ethanol is used. (Source: HEI, 1996.)

Pollutant	Potential Health Effects	Fuels Containing Oxygenates Compared with Conventional Gasoline			
		MTBE Oxyfuel	Ethanol Oxyfuel	MTBE RFG	Ethanol RFG
Oxygenates					
MTBE	• Symptoms (headache, eye irritation, disorientation) • Neurotoxicity • Cancer in animals	↑	0	↑	0
Ethanol	• Effects unlikely when inhaled at low levels (cancer and developmental effects seen with high-level ingestion exposure)	0	↑	0	↑
Air Toxics					
Formaldehyde	• Irritation • Cancer (probable human carcinogen)	↑	0	↑	0
Acetaldehyde	• Cancer (probable human carcinogen)	0	↑	0	↑
Benzene	• Cancer (known human carcinogen) • Developmental effects	↓	↓	↓	↓
1,3-Butadiene	• Cancer (probable human carcinogen)	0?	0?	↓	↓
Carbon Monoxide	• Myocardial ischemia (including angina) during exercise • Decreased exercise capacity	↓	↓	↓	↓
Ozone	• Respiratory symptoms • Lung function decrements • Decreased exercise capacity • Chronic lung injury?	0	0	↓?	↓?

The risk-assessment methodology adopted in the HEI report is largely qualitative. Although the authors present much of the relevant toxicology/emissions and exposures data in quantitative terms, they make no attempt to estimate the ranges of exposure and risk changes associated with the use of oxygenates in gasoline. Rather, they present the likely directions of change (oxygenated fuels compared with conventional fuels) for the substances listed in Table 7.1 and then provide the following overall assessment.

Based on its review of existing evidence on the exposure to and health effects of oxygenates used in gasoline, the HEI Oxygenates Evaluation Committee drew the following conclusions about the oxygenates themselves.

- Introducing oxygenates into gasoline to reduce CO emissions has increased exposure to MTBE for the general public during brief higher-level exposures while refueling and during more sustained but lower-level exposures while driving, and for service station employees during higher-level exposures over entire work shifts. These exposures can occur by both inhalation and skin contact. Workers who handle or transport neat MTBE can experience significantly higher average inhalation exposure levels than people in other situations. Few data on exposure to other oxygenates have been gathered.
- MTBE has been measured in some underground water; its presence in water may result in exposure by ingestion or skin contact should water supplies become contaminated.
- The potential health effects from exposure to gasoline containing MTBE include (1) headaches, nausea, and sensory irritation in some, possibly sensitive, individuals, based on reports after exposure to oxygenates; (2) acute, reversible neurotoxic effects, based on changes in motor activity in rats at high exposure levels; and (3) cancer, based on increases in the frequency of tumors at multiple organ sites in rats and mice at high exposure levels. Although questions persist about how to interpret each of these observed effects, they nevertheless point to a potential human health risk.

- The health effects from exposure to ethanol by ingesting moderate to large quantities have been extensively investigated. Under these conditions, ethanol can increase the risks of certain cancers, adversely affect the developing embryo, produce neurotoxic effects, and cause various other types of damage. However, it is unlikely that such effects would occur at the very low ambient levels to which most people are exposed by inhalation.
- Potential health effects from exposure to other oxygenates are not known and require investigation if their use in fuels is to be widespread.

In addition to these conclusions about the oxygenates themselves, after qualitatively assessing the health effects of gasoline and motor vehicle emissions with and without oxygenates, the Oxygenates Evaluation Committee has come to the following conclusions about gasoline containing oxygenates.

- The potential health effects of exposure to components of conventional gasoline (without oxygenates) include short-term and cancer effects similar to those that could result from exposure to gasoline containing oxygenates.
- Adding oxygenates to gasoline can reduce the emission of CO and benzene from motor vehicles, and thereby potentially lower certain risks to members of the population. At the same time, using oxygenates increases exposure to aldehydes, which are carcinogenic in animals, and to the oxygenates themselves.
- Adding oxygenates is unlikely to substantially increase the health risks associated with fuel used in motor vehicles; hence, the potential health risks of oxygenates are not sufficient to warrant an immediate reduction in oxygenate use at this time. However, a number of important questions need to be answered if these substances are to continue in widespread use over the long term.

In addition to its conclusions about possible health effects, the Oxygenates Evaluation Committee noted a general lesson to be learned from introducing oxygenates to the general public. Although it is not possible to have complete information about a substance before it is used, the diverse experiences after introducing oxygenated fuels argue strongly that any future new use of a substance should (1) be preceded by a sufficiently comprehensive research and testing program (including mechanistic and human studies), and (2) be accompanied by rigorous exposure assessment and epidemiologic studies.

OSTP REPORT

SCOPE

The OSTP report is, by design, far more limited in scope than the HEI report. It focuses entirely on MTBE. It makes no attempt to evaluate the risks of MTBE-containing fuels in relationship to non oxygenated fuels. In effect, the OSTP report is devoted entirely to the possible health risks associated with MTBE exposures resulting from its use in gasoline, in isolation from all other risks.

DATA, METHODOLOGY, AND CONCLUSIONS

The OSTP report presents a review and analysis of available data concerning human exposures to MTBE and the compound's health effects as reported in the currently available human and animal database. Data on the compound's principal metabolites, tertiary-butyl alcohol and formaldehyde, are also reviewed.

With respect to acute health effects, The OSTP report contains a review of available data, but concludes that "the available scientific evidence . . . was considered insufficient to develop estimates of effect at different exposure levels." The report does, however, present estimates of carcinogenic risks associated with MTBE exposures, using the quantitative methodology ordinarily used by EPA. Thus, upper-bound estimates of cancer risks per unit of lifetime average exposure were developed using the data from three animal studies (those described earlier by the committee) and the linearized multistage model. These cancer-potency estimates are presented in Table 7.2, taken from the OSTP report.

These potency estimates are combined with estimates of potential population exposures to MTBE (presented in Chapter 2 of the OSTP report and discussed by our committee) to yield estimates of upper-bound, excess lifetime cancer risks from inhalation of MTBE. These estimates are presented in Table 7.3, taken from the OSTP report.

The principal assumptions underlying the results contained in Table 7.3 are presented in the report's text. Note that, in addition to upper-bound estimates, the authors included so-called maximum likelihood estimates (MLEs) of risk; the footnote to the table explains why the MLEs are not considered reliable.

According to the OSTP report, the greatest risks are those based on the assumption that data on lymphomas and leukemias, obtained in an oral-gavage study in rats (Belpoggi et al., 1995), are predictive of human risk. Under this assumption, service-station attendants may incur an extra lifetime risk from MTBE exposure as high as 1 in 2,000, and certain members of the general population may incur risks as high as 1 in 10,000; as is ordinarily the case, the report emphasizes that the actual risks are not likely to be greater than these upper-bound estimates, could be less, and may even be zero. As noted, no analysis of how these risks compare with risks from unoxygenated gasoline is presented.

Although the OSTP report presents all the available human and animal data concerning the toxicity of MTBE, only acute health effects and possible carcinogenic risks are presented in the final chapter, on risk characterization. The authors present EPA's inhalation reference concentration (RfC) for MTBE, derived with standard regulatory methods and defined by EPA as "an estimate (with uncertainty spanning about an order of magnitude) of a continuous inhalation exposure level for the human population (including sensitive subpopulations) that is likely to be without

Table 7.2 Cancer potency estimates for MTBE based on tumor data from studies in rats and mice.[a] (Source: OSTP, 1996.)

Species	Tumor site	Exposure route	Upper bound unit cancer risk	ED10
Mouse	Liver	Inhalation	6×10^{-4} per ppm[b] 2×10^{-7} per $\mu g/m^3$	460 ppm 480 mg/kg/day
Rat	Kidney	Inhalation	6×10^{-4} per ppm[b] 2×10^{-7} per $\mu g/m^3$	330 ppm 350 mg/kg/day
Rat	Lymphoma/Leukemia	Oral	4×10^{-3} per mg/kg/day	38 mg/kg/day

[a]The cancer potency estimates shown in this table were calculated using the linearized multistage model (Jinot, USEPA, Appendix A) and mouse liver tumor data from the 18 month inhalation study (Burleigh-Flayer et al., 1992), rat kidney tumor data from the 2-year inhalation study (Chun et al., 1992), or lymphoma/leukemia data from the 2-year oral exposure study (Belpoggi et al., 1995). For inhalation exposures, human equivalent daily doses were calculated by adjusting animal exposures of 6 hours/day, 5 days/week to 24 hours/day for 70 years and assuming ppm equivalence between species. For oral exposures, human equivalent doses were calculated by adjusting for exposure once/day, 4 days/week, and applying a surface area correction of (body weight)^{2,3}.

[b]For comparison, estimated upper bound unit cancer risks for fully vaporized conventional gasoline are noted: 2×10^{-3} per ppm based on induction of liver tumors in mice and 4×10^{-3} per ppm based on induction of kidney tumors in rats (EPA, 1987).

Table 7.3 Estimated upper-bound excess inhalation cancer risks and maximum likelihood estimates (MLE) of excess cancer risk based on reasonable worst case time-weighted lifetime (70 years) exposure estimates to MTBE and on cancer potency estimates derived from carcinogenicity data for MTBE in rats and mice. (Source: OSTP, 1996.)[a,b]

Time-weighted Average Lifetime Exposure Estimate (ppm)	Mouse Liver Tumors		Rat Kidney Tumors		Rat Lymphomas/Leukemia[a]	
	Upper-bound excess cancer risk	MLE excess cancer risk	Upper-bound excess cancer risk	MLE excess cancer risk	Upper-bound excess cancer risk	MLE excess cancer risk
4-month oxyfuel season 0.014	8×10^{-6}	5×10^{-16}	9×10^{-6}	2×10^{-10}	7×10^{-5}	4×10^{-5}
6-month oxyfuel season 0.019	1×10^{-5}	1×10^{-15}	1×10^{-5}	4×10^{-10}	9×10^{-5}	6×10^{-5}
6-month oxyfuel and 6-month reformulated gasoline 0.029	2×10^{-5}	5×10^{-15}	2×10^{-5}	8×10^{-10}	1×10^{-4}	8×10^{-5}
Service station attendants 0.10	6×10^{-5}	2×10^{-13}	6×10^{-5}	1×10^{-8}	5×10^{-4}	3×10^{-4}

The actual risks are likely to be somewhat lower than the upper bound calculated risks and could even be nearly zero. The MLE values are highly sensitive to small changes in the tumor incidence data. For example, if the renal tubule adenoma in the control male rat had been present instead in the low exposure group, the MLE values based on the rat kidney tumor data would have been more than 3-4 orders of magnitude lower than the MLE values shown in this table (see Appendix A). In this example, the estimated upper-bound excess cancer risk values change by less than 50%. Because of their instability, the MLEs are not considered reliable.

[a]These cancer risk estimates are derived using a cancer potency estimate from the gavage carcinogenicity study of MTBE in rats with the assumption that the observed tumor response in gavage-treated animals infers an inhalation cancer risk.

[b]Exposure for service station attendants include an 8-hour time-weighted average occupational exposure (5 days per week for 40 years working lifetime) of 0.6 ppm during the 6-month oxygenated fuel season and 0.44 ppm during the 6-month reformulated gasoline season.

appreciable risk for deleterious noncancer effects during a lifetime." EPA's RfC of 3 mg/m^3 (0.83 ppm) was based primarily on findings of increased absolute and relative liver weights and several other effects in rats exposed chronically to MTBE by inhalation.

The OSTP report appears to endorse this RfC as a guide to risk assessments for the documented noncancer health effects of MTBE and concludes that the "reasonable worst-case annual average daily exposure estimate (0.019 ppm)" is well below the protective value. The OSTP report does not deal with the ingestion associated with MTBE, as might exist from consumption of contaminated water.

In addition to its characterization of MTBE's cancer risks, the OSTP report concludes that "it is not known whether the cancer risk of oxygenated gasoline containing MTBE is substantially different from the cancer risk of conventional gasolines." The report also notes that "data were generally inadequate to evaluate the health risks of oxygenates other than MTBE."

GENERAL COMPARISON OF THE TWO REPORTS WITH RESPECT TO RISK ISSUES

It is obvious that the two reports differ substantially regarding scope and methods for risk assessment. The OSTP report, in part because of its stated objective, is very much narrower in scope than the HEI report; but it also states that the type of comparative evaluation undertaken by HEI is not possible with currently available data. HEI's report attempts a comparative evaluation of risk but stops short of any attempt to deal quantitatively with risk.

Although neither report points to a substantial health risk associated with MTBE exposures, it is difficult to compare the two, because of their substantially different scopes, methods, and approaches to scientific uncertainty. It must be noted, however, that the OSTP report describes (albeit with appropriate qualifications) substantial cancer risks associated with certain MTBE exposure scenarios (Table 7.3) but includes no serious attempt to explain their possible public-health importance. Such results, uninterpreted, could be considered at odds with those presented by HEI. (As previously discussed, this committee recommends that risk estimates based on the leukemia-lymphoma findings in the animal study be rejected until further investigation of the results is undertaken.)

Further comparisons of the two reports with respect to their approaches to the risk questions associated with fuel oxygenation would not seem useful here. Rather, we chose to emphasize how the risk-assessment portion of the government's comprehensive evaluation of oxygenated fuels might derive the greatest benefits by recommending ways in which their analyses might be improved.

EVALUATION OF EXISTING ASSESSMENTS AND RECOMMENDATIONS FOR FINAL ASSESSMENT

GENERAL COMMENTS ON RISK ASSESSMENT AND TREATMENT OF UNCERTAINTIES

Both reports appropriately emphasize the substantial uncertainties associated with the problem of developing a comprehensive, comparative assessment of the health risks associated with oxygenated fuels. Indeed, it is clear that the database for certain oxygenates is so meager that no useful analysis can be undertaken. Moreover, the full range of health effects and risks associated with the mixtures of compounds to which populations are exposed through the use of nonoxygenated fuels are not established, and there are, as yet, no data on the health effects of the corresponding mixtures that arise when oxygenated fuels are used (we refer to the mixtures that arise from evaporation, those from combustion, and those from migration into water supplies, which differ from each other). It might be argued that until data are available on these mixtures, no attempt can or should be made to evaluate human health risks.

Such an argument fails on at least two grounds. First, although comprehensive data on the above mentioned mixtures are not available, a substantial amount of data do exist for selected components of those mixtures that are known to present potential health risks. Second, the argument misunderstands the purpose of risk assessment, as elucidated by the National Research Council (NRC 1983, 1994). Risk assessments are undertaken to describe the current state of knowledge regarding the subject being investigated and should include a through description of the uncertainties in that knowledge. Even a risk assessment performed with uncertain data is justified as long as the uncertainties are thoroughly characterized and stated. The risk manager, reviewing the assessor's work, may decide not to use it for decision-making if the uncertainties are judged to be too large, but the assessment itself should avoid such a policy judgment (NRC 1994).

The assessment contained in the HEI report comes close to the type of evaluation the committee thinks is most useful for decision-making regarding the risks associated with oxygenated fuels. It includes an assessment of the risks associated with such fuels, to the extent that those risks are represented by the specific chemicals selected for evaluation and relative to the risks associated with conventional fuels. It includes consideration of the benefits associated with the use of oxygenated fuels. Potential risks associated with water contamination are also described. All these risks are placed in perspective, and the major uncertainties associated with the assessment are well described. We suggest that the HEI report be used as the framework and database (with appropriate additions and reinterpretations, as described above by the committee and in previous chapters) for the comprehensive government risk assessment. In the subsections that follow, we describe the aspects in this report that we think need to be refined for this purpose.

The OSTP report seems far too limited in scope to form the basis for a comprehensive, comparative risk assessment. Although some features of the report are useful (see below), its failure to deal with any of the potential risk issues, on the argument that data are too limited, seems to this committee unsupportable. (The scope of the report, while stated to be narrowly focused on MTBE alone, apparently did not restrain the authors from reaching conclusions regarding the inability to deal with risk comparisons.) The OSTP report fails to present well-documented reasons to support its view regarding the inadequacy of the database for risk assessment (except for the obvious cases of oxygenates on which virtually no data exist).

The other advantage that derives from undertaking a risk assessment is that data and knowledge gaps that are most useful for improving the risk estimates are clearly revealed. The HEI report, because it attempts to be comprehensive, is quite effective in revealing critical research needs.

SUGGESTED REFINEMENTS AND IMPROVEMENTS

Although we recommend that the HEI report be taken as the framework and starting point for the comprehensive risk assessment, we propose that greater effort be made to provide some indication of the magnitudes of the health risks that are said to increase and decrease (relative to conventional fuels) with the use of oxygenates (Table 7.1). We recognize that, regarding the toxicities and dose-response characteristics of the compounds included in the risk analysis, the HEI report is correct with respect to the uncertainties associated with the selection of specific values for human risk assessment. Nevertheless, there are substantial quantitative toxicity data on each of the chemicals, and there is no reason why the federal government cannot adopt reference concentrations, reference doses, and cancer potency factors that have been prepared or developed for all the chemicals using federal guidelines for risk assessment. It is recognized that such values are based in part on science-policy choices (NRC 1994), but, as long as their limitations are appropriately noted, there is no reason why they should not be used for risk assessment. Indeed, federal government agencies regularly issue assessments involving use of such toxicity values. Thus, while the scientists preparing the HEI report may have felt reluctance in describing such measures of toxic risk, we see no reason why the federal government cannot use its standard procedures for such toxicity assessments.

With respect to the exposures described by HEI and their changes following introduction of oxygenates, our comments are similar. It is not necessary to have perfect knowledge regarding exposure and exposure changes; in each case it should be possible to offer some view of at least the likely ranges of exposure to the selected chemicals. The OSTP report, for example, provides some information on the ranges of short-term and long-term MTBE exposures that may be experienced by certain segments of the population. Such an analysis was provided by HEI but was not used in any explicit way in the risk assessment.

Without some attempt to provide an indication of the magnitudes of the risks associated with the potential constituents of oxygenated and nonoxygenated fuels, it is difficult to accept the final conclusions of the HEI report. This is not to say that the assessment should be dominated by quantitative estimates, but only that some attempt to quantify risk ranges should be made. If the ranges of plausible risk increases and decreases overlap substantially, then this would be a reasonable basis for the conclusions reached in the HEI report.

We thus urge the adoption of the scope and framework for risk assessment contained in the HEI report, with greater attention to presentation of quantitative results. Uncertainties in the data should also be carefully described.

SPECIFIC COMMENTS

With respect to some of the specific approaches to risk characterization offered in the two reports, we have the following observations and recommendations.

With respect to noncancer health effects resulting from MTBE inhalation, the EPA's reference concentration (RfC) of 3 mg/m^3 (0.8 ppm) seems appropriate. It is based on the application of standard uncertainty factors and interspecies adjustments to a well-defined NOAEL from a chronic-inhalation study.

Until the issues concerning the possible effects in humans following short-term exposures to MTBE-containing gasolines are clarified, it should be made clear that EPA's RfC may not be appropriate to assess risks of such effects in humans.

No discussion is presented in either report of the need for and value of a reference dose of MTBE—a daily oral intake that is expected to protect against noncancer effects of the compound when it is ingested. Because of the potential for ingestion of MTBE through groundwater that has become contaminated, a close examination of this issue should be undertaken. Several years ago, EPA issued a chronic health advisory for MTBE in drinking water of 20 $\mu g/L$. This value was based on data available in 1992, and several default assumptions were used because of large data deficiencies and uncertainties. The "acceptable" intake (dose) implied by the 20 $\mu g/L$ advisory is substantially less than that implied by the current RfC. An attempt should be made to review currently available toxicity data to establish a reference dose for MTBE.

Because of the substantial doubts regarding the reliability of the leukemia and lymphoma findings reported by Belpoggi et al. (1995), these should not be used for human risk assessment at this time. The estimates of potential human risk reported in Table 7.3, based on these findings, should not be used until the data upon which they are based are verified. When the various questions we have raised have been addressed, a decision can be made regarding the use of these results.

The data for MTBE-induced kidney tumors in male rats should not be used for human risk assessment until the recently reported (CIIT) data on the mechanism of action are reviewed and evaluated. If the new data support the view that α_{2u}-globulin nephropathy is involved in the response, as this committee now believes, then this end point should be discounted for human risk assessment.

The data on MTBE-induced tumors in female mice can be used for risk assessment, but the combining of potency estimates from males and females (OSTP report, Appendix A) is not appropriate. The potency estimate only for females should be used. Although not recommended as the sole basis for assessing the risks, the risk assessment should also consider the NOAEL in the mouse study at 3,000 ppm (human exposure equivalent, 540 ppm) and the margin of exposure separating actual human exposures from this value.

The maximum-likelihood estimates of risk presented in Table 7.3 serve no useful purpose and should be discarded. For reasons stated in the footnote to the table, these estimates are not considered reliable. The committee sees no value in presenting such estimates.

CONCLUSIONS

- The committee has made a number of recommendations for refinements and improvements in the assessment of potential human health risks associated with prolonged exposures to gasoline containing MTBE and for assessment of the comparative risks associated with oxygenated and nonoxygenated fuels. Until these recommendations are acted upon, no definitive statement can be made regarding these health-risk issues. Based on the available analyses, however, it does not appear that MTBE exposures resulting from the use of oxygenated fuels are likely to pose a substantial human health risk.

- It appears that MTBE-containing fuels do not pose health risks substantially different from those associated with nonoxygenated fuels, but this conclusion is less well established and should become the centerpiece for the government's comprehensive assessment.

References

Anderson, H.A., L. Hanrahan, J. Goldring, B. Delaney. 1995a. An Investigation of Health Concerns Attributed to Reformulated Gasoline Use in Southeastern Wisconsin. Final Report, May 30, 1995. Wisconsin Department of Health and Social Services, Milwaukee, WI.

Anderson, H.A., L. Hanrahan, J. Goldring, B. Delaney. 1995b. An Investigation of Health Concerns Attributed to Reformulated Gasoline Use in Southeastern Wisconsin. Phase 2: Telephone Registered Health Concerns. Final Report, September 18, 1995. Wisconsin Department of Health and Social Services, Milwaukee, WI.

Anderson, L.G., P. Wolfe, R.A. Barrell, J.A. Lanning. 1995. The effects of oxygenated fuels on the atmospheric concentrations of carbon monoxide and aldehydes in Colorado in Alternative Fuels and the Environment, F.S. Sterrett, ed. Boca Raton: Lewis Publishers.

Anderson, L.G., P. Wolfe, J.A. Lanning. 1993. The Effects of Oxygenated Fuels on Carbon Monoxide and Aldehydes in Denver's Ambient Air. Proceedings of the Conference on MTBE and Other Oxygenates: A Research Update. Falls Church, VA, July 26-28, 1993. Report no. EPA/600/R-95/134.

Auto/Oil AQIRP. 1991. Technical Bulletin No. 6: Emissions Results of Oxygenated Gasolines and Changes in RVP, Auto/Oil Air Quality Improvement Research Program.

Baker, E., R. Letz, A. Fidler, S. Shalat, D. Plantamura, M. Lyndon. 1985 A computer-based neurobehavioral evaluation system for occupational and environmental epidemiology: methodology and validation studies. Neurobehav. Toxicol. Teratol. 7:369-377.

Barsky, A.J., G. Wyshak, and G.L. Klerman. 1990. The somatosensory amplification scale and its relationship to hypochondriasis. J. Psychiatr. Res. 24(4):323-334.

Beaton, S.P., G.A. Bishop, Y. Zhang, L.L. Ashbaugh, D.R. Lawson, D.H. Stedman. 1995. On-road vehicle emissions: regulations, costs, and benefits. Science. 268:991-993.

Beller, M., and J. Middaugh. 1992. Potential Illness Due to Exposure to Oxygenated Fuels. Fairbanks, Alaska. State of Alaska, Department of Health and Social Services, December 11, 1992.

Beller, M., M. Schloss, and J. Middaugh. 1992. Evaluation of Health Effects from Exposure to Oxygenated Fuel in Fairbanks, Alaska. Bulletin Number 26. Department of Health and Social Services, Anchorage, AK.

Belpoggi, F., M. Soffritti, C. Maltoni. 1995. Methyl-tertiary-butyl ether (MTBE)—a gasoline additive—causes testicular and lymphohaematopoietic cancers in rats. Toxicol. Ind. Health 11:119–149.

Bishop, G.A., D.H. Stedman. 1989. Oxygenated Fuels, a Remote Sensing Evaluation. SAE Technical Paper Series No. 891116.

Bishop, G.A., D.H. Stedman. 1990. On-road carbon monoxide emission measurement comparisons for the 1988-1989 Colorado oxy-fuels program. Environ. Sci. Technol. 24:843-847.

Cain, K.C. and N.E. Breslow. 1988. Logistic regression analysis and efficient design for two-stage studies. Am. J. Epidemiol. 128:1198-206.

Cain W.S., B.P. Leaderer, G.L. Ginsberg, L.S. Andrews, J.E. Cometto-Muñiz, J.F. Gent, M. Buck, L.G. Berglund, V. Mohsenin, E. Monahan, S. Kjaergaard. 1996. Acute exposure to low-level methyl tertiary-butyl ether (MTBE): human reactions and pharmacokinetic responses. Inhalation Toxicol 8:21–48.

California I/M Review Committee. 1993. Evaluation of the California Smog Check Program and Recommendations for Program Improvements. Fourth Report to the Legislature.

CDC (Centers for Disease Control and Prevention). 1993c. An Investigation of Exposure to MTBE and Gasoline among Motorists and Exposed Workers in Albany, New York. Centers for Disease Control and Prevention, Atlanta, GA.

CDC (Centers for Disease Control and Prevention). 1993b. An Investigation of Exposure to Methyl Tertiary Butyl Ether in Oxygenated Fuel in Fairbanks, Alaska. Centers for Disease Control and Prevention, Atlanta, GA.

CDC (Centers for Disease Control and Prevention). 1993a. An Investigation of Exposure to Methyl Tertiary Butyl Ether among Motorists and Exposed Workers in Stamford, Connecticut. Centers for Disease Control and Prevention, Atlanta, GA.

CDC (Centers for Disease Control and Prevention). 1996. CDC white paper. Memorandum signed by Richard J. Jackson and directed to the Interagency Oxygenated Fuels Assessment Steering Committee, dated March 12, 1996 in Interagency Oxygenated Fuel Assessment. 1996. Office of Science and Technology Policy (OSTP) through the Committee on Environment and Natural Resources (CENR) of the President's National Science and Technology Council (NSTC).

Chandler B., and J. Middaugh. 1992. Potential Illness Due to Exposure to Oxygenated Fuels. Anchorage, Alaska. Bulletin Number 1 (Middaugh J, ed.). Department of Health and Social Services, Anchorage, AK.

Cornelius, W.L. 1995. Effects of North Carolina's Oxygenated Fuel Program on Ambient Carbon Monoxide Concentration. Report to the North Carolina Department of Environment, Health, and Natural Resources Air Quality Section.

Daniel, R.A. 1995. Intrinsic Bioremediation of BTEX and MTBE: Field, Laboratory and Computer Modeling Studies.Thesis submitted to Graduate Faculty of North Carolina State University, Department of Civil Engineering, Raleigh, North Carolina.

Davidson, J.M. 1995. Fate and Transport of MTBE—The Latest Data. Pp. 285-301 in the Proceedings of the Petroleum Hydrocarbons and Organic Chemicals in Ground Water: Prevention, Detection, and Remediation Conference, November 29 - December 1,1995. Houston, Texas.

Dolislager, L.J. 1993. Did the wintertime oxygenated fuels program reduce carbon monoxide concentrations in California? Presented at the Tenth International Symposium on Alcohol Fuels, November 4–10, 1993, Colorado Springs, CO.

Egeland, G.M., and D. Ingle. 1995. Ethanol-Blended Fuel in Anchorage, 1994–1995: Residents May Smell the Difference, but Have No Health Complaints. Bulletin Number 19. Department of Health and Social Services, Anchorage, AK.

Fiedler, N., S.N. Mohr, K. Kelly-McNeil, H.M. Kipen. 1994. Response of sensitive groups to MTBE. Inhalation Toxicol. 6:539–552.

Gordian, M.E., M.D. Huelsman, M.L. Brecht, D.G. Fisher. 1995. Health effects of methyl teriary butyl ether (MTBE) in gasoline in Alaska. Alaska Med. 37:101–103.

Gorse, R.A.,Jr., J.D. Benson, V.R. Burns, A.M. Hochhauser, W.J. Koehl, L.J. Painter, R.M. Reuter, and B.H. Rippon. 1991. Toxic Air Pollutant Vehicle Exhaust Emissions with Reformulated Gasolines. SAE Technical Paper Number 912324. Society of Automotive Engineers, Warrendale, PA.

HEI (Health Effects Institute). 1996. The Potential Health Effects of Oxygenates Added to Gasoline. A Review of the Current Literature. A Special Report of the Institute's Oxygenates Evaluation Committee. Health Effects Institute, Cambridge, MA. in Interagency Oxygenated Fuel Assessment. 1996. Office of Science and Technology Policy (OSTP) through the Committee on Environment and Natural Resources (CENR) of the President's National Science and Technology Council (NSTC).

Heil, C.L. 1993. Assessment of the Anchorage Oxygenated Fuels Program on Ambient Carbon Monoxide Concentrations. Master of Environmental Quality Science Thesis, School of Engineering, University of Alaska, Anchorage.

Heywood, J.B. 1988. Internal Combustion Engine Fundamentals. New York: McGraw-Hill.

Hirao, O. And R.K. Pefley. 1988. Present and Future Automotive Fuels: Performance and Exhaust Clarification. New York: John Wiley and Sons.

Hood, J., R. Farina. 1995. Emissions from Light Duty Vehicles Operating on Oxygenated Fuels at Low Ambient Temperatures: A Review of

Published Studies. SAE Draft for Fuels and Lubricants Meeting in Toronto, SAE Paper No. 952403.

Howard, C.J., A. Russell, R. Atkinson, and J. Calvert. 1996. Air Quality Benefits of the Winter Oxyfuel Program in Interagency Oxygenated Fuel Assessmen t. 1996. Office of Science and Technology Policy (OSTP) through the Committee on Environment and Natural Resources (CENR) of the President's National Science and Technology Council (NSTC).

Hubbard, C.E., J.F. Barker, S.F. O'Hannesin, M. Vandegriendt, and R.W. Gillham. 1994. Transport and fate of dissolved methanol, methyl-tertiary-butyl-ether, and monoaromatic hydrocarbons in a shallow sand aquifer. American Petroleum Institute Publication No. 4601: Appendix H, Institute for Groundwater Research, Waterloo, Ontario, Canada, University of Waterloo.

Ingalls, M.N., L.R. Smith, and R.E. Kirksey. 1989. Measurement of On-road Vehicle Emission Factors in the California South Coast Air Basin. Volume 1, Regulated Emissions. Final Report. Prepared by Southwest Research Institute for The Coordinating Research Council, Inc. Atlanta, GA. June 1989.

Interagency Oxygenated Fuels Assessment Ch. 5. 1996. Fuel Economy and Engine Performance Issues in Interagency Oxygenated Fuel Assessment. 1996. Office of Science and Technology Policy (OSTP) through the Committee on Environment and Natural Resources (CENR) of the President's National Science and Technology Council (NSTC).

Johanson, G., A. Nihlen, A. Lof. 1995. Disposition and acute effects of inhaled MTBE and ETBE in male volunteers. Presented at the International Congress of Toxicology-VII; 1995 July; Seattle, WA.

Johanson G., A. Nihlen, and A. Lof. 1995. Toxicokinetics and acute effects of MTBE and ETBE in male volunteers. Toxicol. Lett. 82-83:713-718.

Keislar, R.E., J.L. Bowen, E.M. Fujita, D.R. Lawson. 1995. Effect of Oxygenated Fuels on Ambient Carbon Monoxide Concentrations in Provo, Utah. Final Report prepared by the Desert Research Institute for Geneva Steel Company.

Keislar, R.E., J.L. Bowen, E.M. Fujita, D.R. Lawson, and W.R. Pierson.

1996. Effect of Oxygenated Fuels on Ambient Carbon Monoxide Concentrations in Provo, Utah, USA. Paper presented at 3rd International Symposium on Traffic Induced Air Pollution-Emissions, Impact and Air Quality, Graz, Austria, 29-30 April 1996.

Kirchstetter, T.W., B.C. Singer, R.A. Harley, G.R. Kendall, W. Chan. 1996. Impact of oxygenated gasoline use on California light-duty vehicle emissions. Environ. Sci. Technol. 30:661-670.

Knepper, J.C., W.J. Koehl, J.D. Benson, V.R. Burns, R.A. Gorse Jr., A.M. Hochhauser, W.R. Leppard, L.A. Rapp, R.M. Reuter. 1993. Fuel Effects in Auto/oil High Emitting Vehicles. SAE Technical Paper Series No. 930137.

Koren, H.S., D.E. Graham, R.B. Devlin. 1992. Exposure of humans to a volatile organic mixture. III. Inflammatory response. Arch. Environ. Health 47:39-44.

Kummer, J.T. 1980. Catalysts for automobile emission control. Prog. Energy Combust. Sci. 6:177-199.

Lawson, D.R. 1993. "Passing the test" - human behavior and California's smog check program. J. Air Waste Manage. Assoc. 43:1567-1575.

Lawson, D.R., S. Diaz, E.M. Fujita, S.L. Wardenburg, R.E. Keislar, Z. Lu, and D.E. Schorran. 1996. Program for the Use of Remote Sensing Devices to Detect High-emitting Vehicles. Final Report to the South Coast Air Quality Management District, Desert Research Institute.

Livo, K.B. 1995. Overview of Public's Perspective of Health Effects from Oxygenated Gasolines in Colorado. Presented at HEI Workshop on Acute Health Effects on Oxygenates and Oxygenated Gasolines. July 27, 1995, Chicago, IL.

Mannino, D.M., R.A. Etzel. 1996. Are oxygenated fuels effective? An evaluation of ambient carbon monoxide concentrations in 11 western states, 1986 to 1992. J. Air Waste Manage. Assoc. 46:20-24.

Mayotte, S.C., C.E. Lindhjem, V. Rao, M.S. Sklar. 1994. Reformulated Gasoline Effects on Exhaust Emissions: Phase I: Initial Investigation of Oxygenate, Volatility, Distillation and Sulfur Effects. SAE Technical Paper Series No. 941973.

Mayotte, S.C., V. Rao, C.E. Lindhjem, M.S. Sklar. 1994. Reformulated Gasoline Effects on Exhaust Emissions: Phase II: Continued Investigation of the Effects of Fuel Oxygen Content, Oxygenate Type,

Volatility, Sulfur, Olefins and Distillation Parameters. SAE Technical Paper Series No. 941974.

McCoy, M. Jr, J. Abernethy, T. Johnson. 1995. Anecdotal Health-Related Complaint Data Pertaining to Possible Exposures to Methyl Tertiary Butyl Ether (MTBE): 1993 and 1994 Follow-up Survey (1984–1994). API Publication Number 4623. American Petroleum Institute, Washington, DC.

Medlin, J. 1995. MTBE: The headache of clean air. Environ Health Perspect 103:666–670.

Mehlman, M.A. 1995. Dangerous and cancer-causing properties of products and chemicals in the oil refining and petrochemical industry: part XV, health hazards and health risks from oxygenated automobile fuels (MTBE): lessons not heeded. Int. J. Occup. Med. Toxicol. 4:219–236.

Mirvish, S.S., D.D. Weisenburger, S.H. Hinrichs, J. Nickols, C. Hinman. 1994. Effect of catechol and ethanol with and without methylamylnitrosamine on esophageal carcinogenesis in the rat. Carcinogenesis 15:883–887.

Missoula City-County Health Department. 1993. Oxygenated Fuel Data Collection: Missoula Physician Screening. County Health Department, Missoula, MT.

Mohr, S.N., N. Fiedler, C. Weisel, K. Kelly-McNeil. 1994. Health effects of MTBE among New Jersey garage workers. Inhalation Toxicol. 6:553–562.

Mohr, S.N., P. Nigro, N. Fiedler, K.K. McNeil. 1995. Acute symptoms due to MTBE while driving. Epidemiology. 6 (suppl 4):S17.

Moolenaar, R.L., B.J. Hefflin, D.L. Ashley, J.P. Middaugh, R.A. Etzel. 1994. Methyl tertiary butyl ether in human blood after exposure to oxygenated fuel in Fairbanks, Alaska. Arch. Environ. Health 49:402–409.

Most, W.J. 1989. Coordinating Research Council Study of Winter Exhaust Emissions with Gasoline/Oxygenate Blends. SAE Technical Paper Series No. 892091.

Nihlen, A., R. Walinder, A. Lof, and G. Johanson. 1994. Toxicokinetics and irritative effects of methyl tertiary-butyl ether in man [Unpublished poster presentation]. Presented at a meeting of the

International Society for Environmental Epidemiology; Durham, NC.

NRC (National Research Council). 1983. Risk Assessment in the Federal Government: Managing the Process. Washington, DC: National Academy Press. 191 pp.

NRC (National Research Council). 1991a. Rethinking the Ozone Problem in Urban and Regional Air Pollution. Washington, DC: National Academy Press. 489 pp.

NRC (National Research Council). 1991b. Human Exposure Assessment for Airborne Pollutants—Advances and Opportunities. Washington, DC: National Academy Press. 321 pp.

NRC (National Research Council). 1994. Science and Judgment in Risk Assessment. Washington, DC: National Academy Press. 651 pp.

OSTP (Office of Science and Technology Policy). 1996. Interagency Assessment of Potential Health Risks Associated with Oxygenated Gasoline. National Science and Technology Council in Interagency Oxygenated Fuel Assessment. 1996. Office of Science and Technology Policy (OSTP) through the Committee on Environment and Natural Resources (CENR) of the President's National Science and Technology Council (NSTC).

Pierson, W.R., A.W. Gertler, and R.L. Bradow. 1990. Comparison of the SCAQS tunnel study with other on-road vehicle emission data. J. Air Waste Manage. Assoc. 40:1495-1504.

Poet, T.S., J.E. Murphy, and S.J. Borghoff. 1996. In vitro uptake of methyl t-butyl ether (MTBE) in male and female rat kidney homogenate: Solubility and protein interactions. Toxicologist 30(1, pt.2):abstract 1563, page 305.

Prah, J.D., G.M. Goldstein, R. Devlin, D. Otto, D. Ashley, D. House, K.L. Cohen, T. Gerrity. 1994. Sensory, symptomatic, inflammatory, and ocular responses to and the metabolism of methyl tertiary butyl ether in a controlled human exposure experiment. Inhalation Toxicol. 6:521–538.

PRC Environmental Management, Inc. 1992. Final report of the performance audit of Colorado's oxygenated fuels program. Submitted to Colorado State Auditor.

Prescott-Mathews, J.S., D.C. Wolf, B.A. Wong, and S.J. Borghoff. 1996. Methyl tert-butyl ether (MTBE)-induced protein droplet nephropathy and cell proliferation in male F-344 rats. Toxicologist 30(1, pt.2):abstract 1559, page 304.

Raabe, G.K. 1993. American Petroleum Institute Health Complaint Survey. Proceedings of the Conference on MTBE and Other Oxygenates: A Research Update. Falls Church, VA, July 26-28, 1993. Report no. EPA/600/R-95/134.

Rhudy, S.A., M.O. Rodgers, N. Vescio. 1995. Seasonal Measurements of Motor Vehicle Emissions by Remote Sensing: Raleigh, North Carolina Oxygenated Fuels Program. Submitted to J. Air Waste Manage. Assoc., Sept. 20, 1995.

Robert Bosch GmbH. 1986. Automotive Handbook, 2nd Ed. U. Adler, H. Bauer, W. Bazlen, F. Dinkler, and M. Herwerth. eds. Federal Republic of Germany: Robert Bosch GmbH.

Rothman, K.J. 1986. Modern Epidemiology. Boston: Little Brown and Company.

Shepard, S., J. Heiken, J. Fieber. 1995. Analyses of Michigan and Rosemead Remote Sensing Data Sets, Comparison of MOBILE5a Fleet Emissions to Michigan Remote Sensing Data, and Comparison of EMFAC7F Fleet Emissions to Rosemead Remote Sensing Data. Systems Applications International Technical Memorandum.

Squillace, P.J., D.A. Pope, C.V. Price. 1995a. Occurrence of the Gasoline Additive MTBE in Shallow Ground Water in Urban and Agricultural Areas. U.S. Geological Survey, Fact Sheet FS-114-95, 4 pp.

Stephens, R.D., S.H. Cadle. 1991. Remote sensing measurements of carbon monoxide emissions from on-road vehicles. J. Air Waste Manage. Assoc. 41:39-46.

Thomas, J.M., G.L. Clark, M.B. Thomson, P.O. Bedient, H.S. Rifai, and C.H. Ward. 1988. Envrionmental Fate and Attenuation of Gasoline Components in the Subsurface. Final Report. Report No. DR109. American Petroleum Institute, Washington, DC.

Tosteson, T.D. and J.H. Ware. 1990. Designing a logistic regression study using surrogate measures for exposure and outcome. Biometrika. 77:11-21.

U.S. Environmental Protection Agency. 1993. Assessment of Potential Risks of Gasoline Oxygenated with Methyl Tertiary Butyl Ether (MTBE). Office of Research and Development; Washington, DC. Report no. EPA/600/R-93/206.

Vogt, D. 1994. An Evaluation of the Effect of North Carolina's 1992-93 Oxygenated Fuel Program on Ambient Carbon Monoxide Levels in Urban Areas. Special Report No. 80 by the State Center for Health and Environmental Statistics, North Carolina Department of Environmental, Health and Natural Resources.

Vogt, D., T. Anderson, G. Murray. 1994. Summary of the Evaluation of the Effect of North Carolina's Oxygenated Fuel Program on Ambient Carbon Monoxide Levels in Urban Areas. North Carolina Department of Environment, Health, and Natural Resources Report, Raleigh, NC.

Wayne, L.G., and Y. Horie. 1983. Evaluation of ARB's In-use Vehicle Surveillance Program. Final Report to the California Air Resources Board, ARB Contract No. A2-043-32, Pacific Environmental Services, Inc.

White, M.C., C.A. Johnson, D.L. Ashley, T.M. Buchta, D.J. Pelletier. 1995. Exposure to methyl tertiary-butyl ether from oxygenated gasoline in Stamford, Connecticut. Arch. Environ. Health 50:183–189.

Yeh, C.K., J.T. Novak. 1995. The effect of hydrogen peroxide on the degradation of methyl and ethyl tert-butyl ether in soils. Water Environ. Res. 67(5):828-834.

Zhang, Y., D.H. Stedman, G.A. Bishop, P.L. Guenther, S.P. Beaton. 1995. Worldwide on-road vehicle exhaust emissions study by remote sensing. Environ. Sci. Technol. 29:2286-2294.

Zogorski, J.S., A. Morduchowitz, A.L. Baehr, B.J. Bauman, D.L. Conrad, R.T. Drew, N.E. Korte, W.W. Lapham, J.F. Pankow, and E.F. Washington. 1996. Fuel Oxygenates amd Water Quality: Current Understanding of Sources, Occurrence in Natural Waters, Environmental Behavior, Fate and Significance, Office of Science and Technology Policy in Interagency Oxygenated Fuel Assessment. 1996. Office of Science and Technology Policy (OSTP) through the Committee on Environment and Natural Resources (CENR) of the President's National Science and Technology Council (NSTC).

APPENDIX I

DRAFT
For NAS Review Only

PREFACE

The use of oxygenated gasoline was mandated under the Clean Air Act Amendments of 1990 in areas that did not meet the federal ambient air standard for carbon monoxide (CO). Motor vehicle emissions are the primary source of ambient CO levels in most areas. The Clean Air Act requires at least a 2.7% oxygen content for gasoline sold in CO nonattainment areas, and this level of oxygen is typically achieved by the addition of 15% methyl tertiary butyl ether (MTBE) or 7.5% ethanol (by volume). The higher oxygen content of oxygenated gasoline compared to conventional gasoline is intended to lead to a more complete combustion of the gasoline and therefore to reduced tailpipe emissions of CO.

Soon after the oxygenated gasoline program was introduced nationally in the winter of 1992-1993, anecdotal reports of acute health symptoms were received by health authorities in various areas of the country. Such health concerns had not been anticipated but have subsequently focused attention on possible health risks

[1]Preamble and Executive Summary submitted to the NRC by the Office of Science and Technology on March 15, 1996, as part of the draft interagency report.

associated with using oxygenated gasoline. These health concerns have been joined by complaints of reduced fuel economy and engine performance, as well as the detection of low levels of MTBE in some samples of ground water. The reformulated gasoline program, which is intended to reduce motor vehicle emissions that lead to higher ozone levels during the summer months and air toxics year round, and which also makes use of fuel oxygenates, was not specifically examined in this report.

In order to address public concerns and to take full advantage of the extensive expertise across the Federal government, as well as outside experts where appropriate, the U.S. Environmental Protection Agency (EPA) requested the assistance of the Office of Science and Technology Policy (OSTP) through the Committee on Environmental and Natural Resources (CENR) of the President's National Science and Technology Council (NSTC), to coordinate a comprehensive assessment of these issues. Working groups that prepared this evaluation were comprised of technical and scientific experts from across several Federal agencies, as well as representatives from state government, industry, and environmental groups.

This assessment is a scientific state-of-understanding report of the fundamental basis and efficacy of the EPA's winter oxygenated gasoline program. The assessment considers not only health effects, but also air quality, fuel economy and engine performance, and ground water and drinking water quality. The potential health effects of oxygenated gasoline were evaluated in two separate reports, one prepared by an Interagency group of health scientists and the second by the Health Effects Institute (HEI) and panel of experts. Both the Interagency report and the HEI report, as well as a comparison between these two documents, are included. Each of the chapters in this report underwent extensive external peer-review prior to the submission of the entire report for review by the National academy of Sciences (NAS). The findings and comments from the NAS review will be incorporated into this assessment.

An expanded summary of the health effects of oxygenated gasoline will be prepared based on information in the Interagency and HEI reports and the comments from the NAS review.

When the Interagency Steering Committee began this effort, it was their intention that a full risk assessment and cost benefit analysis of using oxygenated gasoline in place of conventional gasoline be included in the report. However, because of serious limitations in the data, the Steering Committee concluded that such an analysis was not possible at this time. Several research needs on oxygenated gasoline were identified that would reduce uncertainties and allow a more thorough assessment of human exposure, health risks and benefits, and environmental effects.

DRAFT EXECUTIVE SUMMARY

Purpose and Background

Oxygenates have been used as octane enhancers in gasoline since the late 1970s, due to the phaseout of lead. During the 1980s, oxygenates came in to wider use as some states implemented oxygenated gasoline programs for the control of carbon monoxide (CO) pollution in cold weather. People with coronary artery disease are particularly sensitive to the adverse effects of this air pollutant. The first winter oxygenated gasoline program in the United States was implemented in Denver, Colorado in 1988. The 1990 Clean Air Act Amendments required the use of oxygenated gasoline in several areas of the country that failed to attain the National Ambient Air Quality Standard (NAAQS) for CO. During the winter months of 1992-1993, many new oxygenated gasoline programs were implemented to increase combustion efficiency in cold weather and thereby reduce CO emissions.

Methyl tertiary butyl ether (MTBE) has become the most widely used motor vehicle oxygenate in the U.S., though in some areas, ethanol is the dominant oxygenate used in motor vehicle fuels. Other fuel oxygenates that are in use or may potentially be used include ethyl tertiary butyl ether (ETBE), tertiary amyl methyl ether (TAME), diisopropyl ether (DIPE), tertiary butyl alcohol (TBA), and methanol. Because of limitations in available data, there is less emphasis in this report on these other oxygenates. The Clean Air Act requires at least 2.7% by weight oxygen content for gasoline sold in CO nonattainment areas, and 15% by volume MTBE or 7.5% by volume ethanol achieve this requirement.

The purpose of this report is to provide a review of the scientific literature on oxygenated fuels and to assess effects of the winter oxygenated fuels program on air quality, water quality, fuel economy and engine performance, and public health. The request from EPA for this assessment was prompted by public complaints of headaches, nausea, and other acute symptoms attributed to wintertime use of oxygenated fuels, as well as complaints of reductions in fuel economy and engine performance.

This report does not specifically examine the reformulated gasoline program which is intended to reduce motor vehicle emissions that lead to higher ozone levels during the summer months and air toxics year round, and which also makes use of fuel oxygenates. The report identifies areas where the data are too limited to make definitive conclusions about the costs, benefits, and risks of using oxygenated gasoline in place of conventional gasoline. Several research needs on oxygenated gasoline were identified that would reduce uncertainties and allow a more thorough assessment of human exposure, health risks and benefits, and environmental effects.

Assessment Findings

Air Quality

- A general decline in urban concentrations of CO over the past twenty years is attributed to stringent EPA mandated vehicle emission standards and improved vehicle emission control technology. The effects of meteorology must be accounted for in assessments of air quality benefits of oxygenated gasoline.
- In some cities with winter oxygenated gasoline programs, a reduction in ambient CO concentrations of about 10% is observed and is attributed to the use of the oxygenate.
- Studies of the effects of fuel oxygenates on vehicle emissions show a consistent reduction of CO emissions at ambient temperatures above about 50°F. At temperatures below 50°F, the magnitude of the reduction is decreased and more uncertain. Some studies show increased CO emission when oxygenated fuels are used at low temperature.
- The EPA MOBILE 5a model appears to overestimate the benefits of oxygenated gasoline on fleetwide CO emissions by a factor of two.
- Oxygenates also reduce total hydrocarbon exhaust emissions. Fuel oxygenates decrease vehicle emissions of air toxics, benzene and 1,3-butadiene, but increase the emissions of aldehydes (acetaldehyde from use of ethanol or ETBE and formaldehyde from use of MTBE).
- The amount of pollutant emissions is smaller in newer technology vehicles (fuel injected and adaptive learning, closed loop three-way catalyst systems) than in older technology vehicles (carbureted and oxidation catalysts). Also, the percentage reductions in CO and hydrocarbon emissions from use of fuel oxygenates are found to be smaller in the newer technology vehicles compared to older technology and higher emitting vehicles.

• Emissions of nitrogen oxides are not changed significantly by low concentrations of fuel oxygenates but some studies indicate increased nitrogen oxide emissions with oxygenate concentrations greater than 2 percent by weight oxygen.
• During the winter season, the oxygenates are not removed rapidly from the urban atmosphere, although some scavenging by precipitation is expected. Consequently, the oxygenates are likely to be dispersed and diluted throughout the troposphere, where they ultimately would be removed by slow photooxidation.

Water Quality

• Releases of gasoline containing oxygenates to the subsurface from storage tanks, pipelines, and refueling facilities provide point sources for entry of high concentrations of fuel oxygenates into ground water. Underground storage tank improvement programs underway by the states and EPA should result in a reduction in the release of gasoline and fuel oxygenates to ground water from these potential point sources.
• Exhaust emissions from vehicles and evaporative losses from gasoline stations and vehicles are sources of oxygenate release to the atmosphere. Because of their ability to persist in the atmosphere for days to weeks and because they will partition into water, fuel oxygenates are expected to occur in precipitation in direct proportion to their concentration in air. Hence, fuel oxygenates in the atmosphere provide a non-point, low concentration source to the hydrologic cycle as a result of the dispersive effect of weather patterns and occurrence in precipitation.
• Volatilization of the alkyl ether oxygenates will occur slowly from ponds and lakes, and from slow-moving and deep streams and rivers; volatilization can be rapid from shallow and fast-moving streams and rivers. Alkyl ether oxygenates are much less biode-

gradable than ethanol or the aromatic hydrocarbon constituents of gasoline and, therefore, will persist longer in ground water. They also adsorb only weakly to soil and aquifer material. Consequently, dissolved alkyl ether oxygenates will move with the ground-water flow and migrate further from a point source of contamination.

- The USEPA draft drinking-water lifetime health advisory for MTBE ranges from 20 to 200 μg/L; a revised health advisory is expected later this year. Health advisories have not been developed for other fuel oxygenates.
- MTBE was detected in 7% of 592 storm-water samples in 16 cities surveyed between 1991-1995. When detected, concentrations ranged from 0.2 to 8.7 μg/L, with a median of 1.5 μg/L. A seasonal pattern of detections was evident, as most of the detectable concentrations occurred during the winter season. MTBE was detected both in cities using oxygenated gasoline to abate CO nonattainment and in cities using MTBE-oxygenated gasoline for octane enhancement.
- At least one detection of MTBE has occurred in ground water in 14 of 33 states surveyed. MTBE was detected in 5% of approximately 1500 wells sampled, with most detections occurring at low (μg/L-level) concentrations in shallow ground water in urban areas.
- Drinking water supplied from ground water has been shown via limited monitoring to be a potential route of human exposure to MTBE. MTBE has been detected in 51 public drinking water systems to date based on limited monitoring in 5 states, however, when detected the concentrations of MTBE were for the most part below the lower limit of the USEPA health advisory. Because of the very limited data set for fuel oxygenates in drinking water, it is not possible to describe MTBE's occurrence in drinking water nationwide nor to characterize human exposure from consumption of contaminated drinking water.

• There is not sufficient data on fuel oxygenates to establish water quality criteria for the protection of aquatic life.
• The presence of MTBE and other alkyl ether oxygenates in ground water does not prevent the application of conventional (active) methods to clean up gasoline releases; however, the cost of remediation involving MTBE will be higher. Also, the use of intrinsic (passive) bioremediation to clean up gasoline releases containing MTBE may be limited because of the difficulty with which MTBE is biodegraded.

Fuel Economy and Engine Performance

• Theoretical predictions based on energy content indicate that reductions in fuel economy resulting from the addition of allowable levels of oxygenates to gasoline should be in the range of only 2-3%. On-road measurements agree with these estimates.
• Automobile engine performance problems due solely to the presence of allowable levels of oxygenates in gasoline are not expected because oxygenated gasolines and nonoxygenated gasolines are manufactured to the same specifications of the American Society for Testing and Materials.

Health Effects

• Complaints of acute health symptoms, such as headaches, nausea, dizziness, and breathing difficulties, were reported in various areas of the country after the introduction of oxygenated gasoline containing MTBE. Limited epidemiological studies and controlled exposure studies conducted to date do not support the contention that MTBE as used in the winter oxygenated fuels program is causing significant increases over background in acute symptoms or illnesses in the general public or workers; however, they do not rule out the possibility that a small percentage of the population may be sensitive to MTBE alone or in gasoline.

• Human exposure data to MTBE are too limited for a quantitative estimate of the full range and distribution of exposures to MTBE among the general population. Less information is available on exposures to oxygenates other than MTBE.

• The assessment found that chronic non-cancer health effects (neurologic, developmental, or reproductive) would not likely occur at an environmental or occupational exposures to MTBE. The observation of acute and reversible neurobehavioral changes in rats exposed to relatively high levels of MTBE is indicative of a neuroactive or possibly neurotoxic effect.

• Current data are too limited to quantitate health benefits of reduced ambient CO from wintertime use of oxygenated fuels.

• Experimental studies indicate that MTBE is carcinogenic in rats and mice at multiple organ sites after oral or inhalation exposure. The mechanisms by which MTBE causes cancer in animals are not well understood. Tertiary butyl alcohol and formaldehyde, the primary metabolites of MTBE biotransformation, are also carcinogenic in animals.

• MTBE has been tested for genotoxicity with generally negative results, whereas formaldehyde is genotoxic in a variety of experimental systems. While there are no studies on the carcinogenicity of MTBE in humans, based on animal data there is sufficient evidence to conclude that MTBE is either possibly or probably a human carcinogen. However, estimates of human risk from MTBE contain large uncertainties in both human exposure and cancer potency.

• The interpretation of any health risks associated with the addition of MTBE to gasoline requires a comparison to the health risks associated with conventional gasoline. The net effects of oxygenated gasoline on emissions and ambient concentrations of air toxics have not been adequately characterized. Consequently, comparative risks between oxygenated and nonoxygenated gasolines have not been established.

- It is not likely that the health effects associated with ingestion of moderate to large quantities of ethanol would occur from inhalation of ethanol at ambient levels to which most people may be exposed from use of ethanol as a fuel oxygenate. Potential health effects from exposure to other oxygenates are not known and require investigation if their use in fuels is to become widespread.